METAL RECOVERY FROM ELECTRONIC WASTE: TWO-STEP BIOLOGICAL AND CHEMICAL LEACHING FOR THE RECOVERY OF COPPER AND GOLD FROM ELECTRONIC WASTE

Arda Işıldar

Joint PhD degree in Environmental Technology

UNIVERSITÉ ━━━━
—PARIS-EST

Docteur de l'Université Paris-Est
Spécialité : Science et Technique de l'Environnement

Dottore di Ricerca in Tecnologie Ambientali

UNESCO-IHE
Institute for Water Education

Degree of Doctor in Environmental Technology

Tesi di Dottorato – Thèse – PhD thesis

Arda Işıldar

Two-step biological and chemical leaching for the recovery of copper and gold from electronic waste

Defended on November 18th, 2016

In front of the PhD committee

Professor Erkan Şahinkaya	Reviewer
Associate Professor Jaco Huisman	Reviewer
Associate Professor Ester van der Voet	Reviewer
Doctor Fritz Holzwarth	Examiner
Professor Piet N. L. Lens	Promotor
Associate Professor Eric D. van Hullebusch	Co-Promotor
Associate Professor Giovanni Esposito	Co-Promotor

Thesis committee:

Thesis Promotor:

Prof. Dr. ir. Piet N. L. Lens

Professor of Biotechnology

UNESCO-IHE Institute for Water Education, Delft, the Netherlands

Thesis Co-promotors:

Dr. Hab. Eric D. van Hullebusch

Hab. Associate Professor of Biogeochemistry

Université Paris-Est Marne-la-Vallée, Paris, France

Dr. Giovanni Esposito

Associate Professor of Sanitary and Environmental Engineering

Università degli Studi di Cassino e del Lazio Meridionale, Cassino, Italy

Other members:

Prof. Erkan Şahinkaya

Professor of Environmental Engineering

Faculty of Bioengineering, Istanbul Medeniyet Üniversitesi, Istanbul, Turkey

Dr. Jaco Huisman

Associate Professor of Design for Sustainability and Product Innovation Management

Faculty of Industrial Design Engineering, TU Delft, Delft, the Netherlands

Dr. Ester van der Voet

Associate Professor of Industrial Ecology

Institute of Environmental Sciences (CML), Leiden University, Leiden, the Netherlands

Dr. Fritz Holzwarth

Rector ad interim of UNESCO-IHE

UNESCO-IHE Institute for Water Education, Delft, the Netherlands

This research was conducted under the auspices of Erasmus Mundus Joint Doctorate Environmental Technologies for Contaminated Solids, Solids and Sediments (ETeCoS[3]) and the Graduate School for Socio-Economic and Natural Sciences of the Environment (SENSE)

CRC Press/Balkema is an imprint of the Taylor & Francis Group, an informa business

Published by:

CRC Press/Balkema

Schipholweg 107C, 2316 XC, Leiden, the Netherlands

Pub.NL@taylorandfrancis.com

www.crcpress.com – www.taylorandfrancis.com

ISBN 978-0-367-08705-0

''Look deeper into the nature, and then you will understand everything better.''

Albert Einstein

Acknowledgments

This PhD work would have not been successfully and joyfully completed without the contributions of many exceptional people, who made this research a life experience rather than a mere research project.

My sincere gratitude goes to my promoter Prof. Piet N.L. Lens who acutely dedicated his time, effort and valuable experience to this particular research project. My academic path and learning journey would not have shaped without his indispensable mentorship, training and guidance.

When it comes to gratitude, my co-promoter Prof. Eric D. van Hullebusch deserves a special reservation, who undeniably formed my professional development and helped me open many new frontiers in my career. Without doubts, Eric will remain a lifelong counsel, an esteemed colleague, and a friend, beyond this PhD project.

The contribution of Dr. Eldon Rene shaped the skeleton of this PhD work and bequeathed virtually all parts of this journey. His outstanding ability to balance professionalism and extracurricular activities made an excellent example of an esteemed researcher and chemical engineer. Dr. Jack van de Vossenberg was a source of inspiration, without whom this work would have not been completed, and a keen, detail-oriented, critical reviewer of scientific findings.

Dr. Venkata Nancharaiah Yarlagadda facilitated the laboratory sessions in the best way any PhD fellow can dream of. Prof. Giovanni Esposito was always promptly available for any issue that may rise, and organized the enjoyable Summer School evenings in Cassino. Thank you, Giovanni, for coordinating the ETeCoS^3 programme.

The data would have not been accurately delivered without the tireless support of Ferdi, Peter Frank, Berend, Fred in Delft, and Chloé and Yoan in Paris. Especially the impressive teamwork of the UNESCO-IHE lab staff showed that a whole was more than the sum of its individual parts.

I had the exceptional privilege to work with excellent M.Sc. students Bienvenu Mizero, Thresa Musongo, Marco Villares, Yaowen Wang, and Nuria Caseres. Neither the data would have been comprehensive, nor the discussion would have been in-depth without their valuable contribution. Special thanks go to Mr. Marco Villares, who not only shaped the course of this

PhD work but also became an indispensable companion in professional and off-work environments.

Many thanks to all the ETeCoS[3] and ABWET colleagues; Chiara, Joy, Rohan, Suthee, Carlos, Susma, Anna, Manivannan, Lucian, Joana, Clement, Soliu, Douglas, Francesco, Ludovico, Kirki, Lea, Shrutika, Suchanya, Anna, who are not only fellow researchers but also friends and a source of a collective aura. The Summer Schools would have not been the same without the spontaneous sangria parties. Friends scattered around cities; Eirini, Iosif, Marco, Chris, Paulo, Mohaned, Alex, Poolad, Paolo, Maria, Sanaz, Ali, Germana, Angélica, Paulo, Maria, Nico, Veronica, Xiaoxia, Alessandra, Ekin, Omar, I celebrate our lifelong friendship built on invaluable memories acquired in Delft, Leiden, Paris and Cassino.

Special thanks go to the colleagues at SIMS Recycling in Eindhoven, and at UNESCO-IHE IT department, colleagues at UNESCO-IHE, and the citizens of city of Delft for kindly providing the WEEE samples. And finally, I would like to address my appreciation to the sponsor of this work, the European Commission, which financially supported the project.

Summary

The well-being of the society depends on a number of metals, including base metals, precious metals and increasingly rare earth elements (REE). The usage of these metals increased in numerous applications, including electrical and electronic equipment (EEE), and their interrupted supply is at stake. There is an increasing interest in the secondary sources of these metals, particularly waste electrical and electronic equipment (WEEE) in order to compensate their potential supply deficit. This PhD thesis demonstrates the advantages and bottlenecks of biological and chemical approaches, as well as the advances and perspectives in the development of sustainable processes for metal recovery from WEEE. Furthermore, a novel process for the recovery of metals from WEEE is described, and a techno-economic assessment is given.

Discarded printed circuit boards (PCB) from personal computers (PC), laptops, mobile phones and telecom servers were studied. Following an extensive literature review, a novel characterization and total metal assay method was introduced and applied to waste board materials. Discarded PCB contained metals in the range of (%, by weight): copper (Cu) 17.6 - 39.0, iron (Fe) 0.7 - 7.5, aluminum (Al) 1.0 - 5.5, nickel (Ni) 0.2 - 1.1, zinc (Zn) 0.3 - 1.2, as well as gold (Au) (in ppm) 21 - 320. In addition, multi-criteria analysis (MCA) using the analytical hierarchical process (AHP) methodology was applied for selection of the best-suited technology. A proof-of-concept for a two-step bioleaching extraction is given, in which 98.4% and 44.0% of the Cu and Au, respectively, were extracted. The two-step extraction procedure was applied to the chemical leaching of metals from PCB. Cu leaching was carried in an acidic oxidative mixture of H_2SO_4 and H_2O_2, whereas Au was leached by $S_2O_3^{2-}$ in a NH_4^+ medium, catalyzed by $CuSO_4$. Under the optimized conditions, 99.2% and 92.2% of Cu and Au, respectively, were extracted from the board material. Selective recovery of Cu from the bioleaching leachate using sulfidic precipitation and electrowinning is studied. Cu was selectively recovered on the cathode electrode at a 50 mA current density in 50 minutes, with a 97.8% efficiency and 65.0% purity. The techno-economic analysis and environmental sustainability assessment of the new technology at an early stage of development was investigated.

Sommario

Il benessere della societá dipende dal numero di metalli, compresi i metalli vili, i metalli preziosi e sempre di piú le terre rare (rare earth elements, REE). L'uso di questi metalli é aumentato in varie applicazioni, anche nelle apparecchiature elettriche ed elettroniche (AEE) e la loro ininterrota provvista é a rischio. Vi é un crescente interesse nelle fonti secondarie di questi metalli, particolarmente nei rifiuti delle apparecchiature elettriche ed elettroniche (RAEE), per compesare il loro potenziale deficit di approvvigionamento. Questa tesi di dottorato dimostra i vantaggi e le difficoltá delle strategie chimiche e biologiche, nonché i progressi e le prospettive dello sviluppo dei processi sostenibili per il recupero dei metalli dai RAEE. Inoltre, viene qui descritto un nuovo processo per il recupero dei metalli dai RAEE e ne viene fornita una valutazione tecnico-economica.

Sono stati investigati circuiti stampati (printed circuit boards, PCB) che vengono scartati da personal computer (PC), computer portatili, telefoni cellulari e server di telecomunicazioni. A seguito di un'ampia revisione letteraria, un nuovo metodo per la caratterizzazione e per il dosaggio dei metalli é stato introdotto e applicato ai materiali di rifiuto da PCB. I PCB scartati contenevano metalli in diversa percentuale di peso (%), compresa tra: 17.6 - 39.0 di rame (Cu), 0.7 - 7.5 di ferro (Fe), 1.0 - 5.5 di alluminio (Al), 0.2 - 1.1 di nichel (Ni), 0.3 - 1.2 di zinco (Zn), cosí come 21 - 320 (in ppm) di oro (Au). Inoltre, per selezionare la tecnologia piú appropriata é stata applicata un'analisi multi-criterio (AMC) utilizzando un processo d'analisi gerarchico (analitycal hierarchical process, AHP). Una verifica teorica é stata eseguita sulla biolisciviazione a due fasi per l'estrazione dei metalli, nella quale sono stai estratti rispettivamente il 98.0% e il 44.0% di Cu e Au. La procedura di estrazione a due fasi é stata applicata alla lisciviazione chimica dei metalli dai PCB. La lisciviazione di Cu é stata effettuata in una miscela acido-ossdativa di H_2SO_4 e H_2O_2, mentre per l'Au é stato usato il $S_2O_3^{2-}$ in una soluzione di coltura con NH_4^+, catalizzato da $CuSO_4$. Nelle condizioni ottimali, il 99.2% di Cu e il 92.2% di Au sono stati estratti dai PCB. E' stato investigato il recupero selettivo di Cu dal percolato di biolisciviazione attraverso la precipitazione con il sulfuro e l'elettroraffinazione. Il Cu é stato selettivamente recuperato al catodo ad una densitá di corrente di 50 mA in 50 minuti, con un'efficienza del 97.8% e il 65.0% di purezza. Sono state effettuate l'analisi tecnico-economica e la valutazione sulla sostenibilitá ambientale di questa nuova tecnologia nuova tecnologia nella sua fase primordiale di sviluppo.

Samenvatting

Het welzijn van onze samenleving is afhankelijk van verscheidene metalen, inclusief onedele metalen, edele metalen, en in toenemende mate zeldzame aardmetalen. Het gebruik van deze metalen is toegenomen in vele toepassingen, inclusief elektrische en elektronische apparatuur (EEA), en hun onderbroken aanvoer staat op het spel. Moderne elektronische apparaten kunnen tot wel 60 metalen bevatten. Er is een toenemende interesse in secundaire bronnen voor deze metalen, vooral afgedankte elektrische en elektronische apparatuur (AEEA), om het mogelijke tekort in aanvoer te compenseren. Dit proefschrift demonstreert de voordelen en knelpunten van biologische en chemische aanpakken, alsmede de vooruitgang en perspectieven in de ontwikkeling van processen voor het herwinnen van metaal uit AEEA. Bovendien wordt een nieuwe methode voor de herwinning van metalen uit AEEA beschreven en een technisch-economische beoordeling gegeven.

Afgedankte printplaten uit persoonlijke computers (PC), laptops, mobiele telefoons, en telecom servers zijn bestudeerd. Na een extensieve literatuurstudie is een nieuwe karakterisatie en totale metaal keuring geïntroduceerd en toegepast op de afgedankte platen. De afgedankte printplaten bevatten de volgende metalen (% van het gewicht): koper (Cu) 17.6 -39.0. ijzer (Fe) 0.7 -7.5, aluminium (Al) 1.0 - 5.5, nikkel (Ni) 0.2 - 1.1, zink (Zn) 0.3 - 1.2, en goud (Au) (in ppm) 21 - 320. Daarnaast is een multicriteria analyse (MCA), met gebruik van de analytische hiërarchisch proces (AHP) methode, toegepast voor de keuze van de meest geschikte technologie. Een 'proof of concept' voor een 'bioleaching' extractie in twee stappen wordt beschreven, waarbij 98.4% van de koper, en 44.0% van het zilver onttrokken werd. De extractieprocedure werd toegepast voor het chemisch uitlogen van metalen uit printplaten. Uitloging van Cu werd uitgevoerd in een zuur, oxidatief mengsel van H_2SO_4 en H_2O_2, terwijl Au werd uitgeloogd door $S_2O_3^{2-}$ in een NH_4^+ medium, gekatalyseerd door $CuSO_4$. Onder geoptimaliseerde omstandigheden werd 99.2% Cu en 92.2% Au uit de printplaat ontrokken. Selectieve herwinning van Cu uit de 'bioleaching' loging met behulp van sulfidische neerslag en electrowinning is bestudeerd. Cu werd selectief herwonnen van de kathode bij een stroom van 50 mA gedurende 50 minuten, met een effectiviteit van 97.8% en een puurheid van 65.0%. De technisch-economische analyse en duurzaamheid van de nieuwe technologie in een vroeg stadium van de ontwikkeling zijn onderzocht.

Résumé

Le bien-être de notre société dépend directement de plusieurs métaux tels que les métaux communs, les métaux précieux et, de plus en plus, les terres rares (TR). L'utilisation de ces métaux s'est développée dans de nombreuses applications, notamment pour les équipements électriques et électroniques (EEE), et leur approvisionnement interrompu est un enjeu majeur. Les appareils électroniques modernes contiennent jusqu'à 60 métaux différents. Il y a un intérêt grandissant pour les sources secondaires de ces métaux, en particulier les déchets d'équipements électriques et électroniques (DEEE), afin de compenser des potentiels manques d'approvisionnement. Cette thèse de doctorat montre les avantages et les inconvénients des approches biologiques et chimiques, ainsi que des avancées et perspectives dans le développement de procédés viables for la récupération des métaux des DEEE. Un nouveau procédé for la récupération des métaux des DEEE est décrit et une évaluation économique est fournie.

Les cartes de circuits imprimés (CCI) des ordinateurs de bureau, des ordinateurs portables, des téléphones mobiles et des serveurs de télécommunications ont été étudiées. Les CCI jetées contenaient des concentration en métaux dans la gamme (% du poids) cuivre (Cu) 17,6 - 39,0 / fer (Fe) 0,7 - 7,5 / aluminium (Al) 1,0 - 5,5 / nickel (Ni) 0,2 - 1,1 / zinc (Zn) 0,3 - 1,2 , ainsi que de l'or (Au) (en ppm) 21 - 320. Une analyse multicritère (AMC) utilisant la méthodologie du processus d'analyse hiérarchique (PAH) a été appliquée pour la sélection de la technologie de récupération des métaux la plus adaptée. Une preuve du concept d'extraction par une double étape de biolixiviation est fournie, dans laquelle 98,4% et 44,0% de cuivre et d'or, respectivement, ont été extrait. Cette procédure d'extraction à deux étapes a aussi été appliquée pour la lixiviation chimique des métaux des CCI. La lixiviation du Cu a été effectuée dans un mélange acide d'H_2SO_4 et d'H_2O_2, alors que l'Au a été extrait par du $S_2O_3^{2-}$ dans un milieu NH_4^+, catalysé par $CuSO_4$. Avec les conditions opératoires optimales, 99,2% et 92,2% de Cu et d'Au, respectivement, ont été extrait de ces matériaux. La récupération sélective du Cu du lixiviat de biolixiviation a été étudiée en utilisant la précipitation sulfurée et l'extraction électrolytique (electrowinning). Le Cu a été récupéré de manière sélective en 50 min sur la cathode à une densité de courant de 50 mA, avec une efficacité de 97,8% et une pureté de 65,0%. L'analyse technico-économique et l'évaluation de la viabilité environnementale de la nouvelle technologie à un stade précoce de développement ont été étudiées.

Contents

Nomenclature

AHP	Analytic hierarchy process
AMD	Acid mine drainage
BAT	Best available technologies
BFRs	Brominated flame retardants
CCD	Central composite design
CSTR	Continous stirred tank reactor
CPU	Central processign unit
CRT	Cathode ray tube
EEE	Electric and electronic equipment
ELCD	European life cycle database
EoL	End-of-life
EPR	Extended producer responsibility
EW	Electrowinning
GDP	Gross domestic product
GHG	Greenhouse gas emissions
HDD	Hard disc drive
ICT	Information and communication technology
ITO	Indium tin oxide
MCA	Multi-criteria assesment
MCDA	Multi-criteria decision analysis
LoC	Loss on comminution
LCA	Life-cycle assessment
LCD	Liquid crystal display
LCI	Life-cycle impact
LCIA	Life-cycle impact assessment
LED	Light emitting diode
LME	London metal exchange
LS	Leachate solution
MLCC	Multi-layer ceramic chip capacitors
NADH	Nicotinamide adenine dinucleotide
NdFeB	Neodymium iron boron

ORP	Oxidation and reduction potential
PAHs	Polycyclic aromatic hydrocarbons
PBB	Polybrominated biphenyls
PBDD/Fs	Polybrominated dibenzo-p-dioxins and dibenzofurans
PBDEs	Polybrominated diphenyl ethers
PCB	Printed circuit board
PCBs	polychlorinated biphenyls
PGM	Platinum group metals
PSD	Particle size distrubution
RoHS	Restriction on the use of hazardous substances
RSM	Response surface methodology
REE	Rare earth elements
SEM	Scanning electron microscope
StEP	Solving the E-waste problem initiative
SRB	Sulfate-reducing bacteria
SS	Synthetic solution
TCI	Total capital investment
TCIC	Total capital investment costs
TIIC	Total indirect investment costs
TPEC	Total purchased equipment costs
TRL	Technology readiness level
TOC	Total operational costs
WEEE	Waste electrical and electronic equipment
WtE	Waste-to-energy
XRD	X-ray diffraction

Chapter 1.
Introduction

1.1 Introduction

1.1.1 Waste electrical and electronic equipment (WEEE) as a secondary source of metals

Global waste electrical and electronic equipment (WEEE) generation was 41.8 million tons (Mt) in 2014, of which 9.5, 7.0 and 6.0 Mt belonged to EU-28, USA and China, respectively (StEP, 2015), and is likely to increase to 50 Mt in 2018 (Baldé et al., 2015). Low lifespan of electronic devices, perpetual innovation in electronics (Ongondo et al., 2015) and affordability of the devices (Wang et al., 2013) resulted in an unprecedented increase of WEEE. Despite the growing awareness and deterring legislation, most of the WEEE is disposed improperly, mostly landfilled (Cucchiella et al., 2015) or otherwise shipped overseas (Ladou and Lovegrove, 2008) to be treated in substandard conditions. Illegal shipping of such waste is a very important problem, currently dealt at an international level. When exported to the developing economies, the costs of WEEE treatment are externalized (Mccann and Wittmann, 2015).

Management of WEEE is of environmental and social concern with global implications due to its hazardous nature. The nature of the production, distribution and disposal of electronic devices include global chains (Breivik et al., 2014). The source of the global WEEE problem has its roots in lack of technologically mature solutions, poor enforcement and high costs of legal operations, and waste being a global commodity in contrast with the regulations (Baird et al., 2014). It is simply cheaper for the end users to ship the waste material overseas. Lack of an effective technical solution, so as to efficiently and selectively recovery metals plays a major role (Lundgren, 2012). In addition to all the hazards originating from WEEE, manufacturing electrical and electronic equipment (EEE) consumes considerable amounts of minerals, particularly metals. Electronics industry is the third largest consumer of gold (Au), responsible of 12% of the global demand, along with 30% for copper (Cu), silver (Ag) and tin (Sn) (Mccann and Wittmann, 2015). More than one million people in 26 countries across Africa, Asia and South America work in gold mining, mostly in unregistered substandard conditions (Schipper et al., 2015).

The rapid increase of EEE production and consequent WEEE generation are reliant on access to a number of raw materials. Many of them are critical due to their limited supply, potential usage in other applications and economic importance (Bakas et al., 2014). The number of materials used in hi-tech products tremendously increased. WEEE is a complex mixture of different materials in various concentrations. Modern devices encompass up to 60 elements, with an increase of complexity with various mixtures of compounds (Bloodworth, 2014). These elements go into the manufacture of microprocessors, circuit boards, displays, and permanent magnets usually in tiny quantities and often in complex alloys (Reck and Graedel, 2012). Discarded printed circuit boards (PCB) are an important secondary source of valuable metals. All EEE contains PCB (Marques, 2013) of various size, type and composition (Duan et al., 2011). These materials are a complex mixture of metals, polymers and ceramics (Yamane et al., 2011).

WEEE contains considerable quantities of valuable metals such as base metals, precious metals and rare earth elements (REE). These 'specialty' metals are used to enable enhanced performance in modern high-tech applications and are collectively termed technology metals (Reck and Graedel, 2012). Typically, a PCB includes very high concentrations of metals such as copper (Cu), iron (Fe), aluminum (Al), and nickel (Ni) along with precious metals such as gold (Au), silver (Ag), platinum (Pt), and palladium (Pd). Metal concentrations of discarded PCB are much higher than those of the natural ores. Metal recovery from discarded WEEE is conventionally carried out by pyrometallurgical and hydrometallurgical methods, which have their own drawbacks and limitations (Cui and Zhang, 2008). The composition of PCB after processing from a WEEE treatment plant is 38.1% ferrous metals, 16.5% non-ferrous metals, and 26.5% plastic, and 18.9% others (Bigum et al., 2012). Precious metals are the main driver of recycling (Hagelüken, 2006), viz. Au has the highest recovery priority; followed by copper (Cu), palladium (Pd), aluminium (Al), tin (Sn), lead (Pb), platinum (Pt), nickel (Ni), zinc (Zn) and silver (Ag) (Wang and Gaustad, 2012). On the other hand, the intrinsic value of non-precious technology (speciality) metals is increasing (Tanskanen, 2013) owing to decreasing concentration of precious metals in PCB (Luda, 2011; Yang et al., 2011).

Informal recycling of WEEE has catastrophic effects on the people and the environment. In Europe, where there is a tradition of preventative legislation and robust policy measures, only 35% (3.3 Mt) of WEEE is reported to be officially collected (Huisman et al., 2015), and the rest is speculated to be exported, treated under substandard conditions, or simply thrown in waste bins (Figure 1-1). The hazards associated with improper WEEE management come twofold: degradation of the environment (Song and Li, 2014) and loss of valuable resources (Oguchi et al., 2013). Despite its toxicity, PCB contains valuable materials that could be recovered to yield both environmental and economic benefits (Kumari et al., 2016; Liang et al., 2010).

Figure 1-1: Informal processing of electronic waste; discarded central processing units (CPU) for recycling (a), electronic waste dumb in Ghana (b), substandard processing in Shanghai, China (c) worker in Guiyu, China (d).

Urban areas are densely populated by obsolete end-of-life (EoL) electronic devices. These waste materials are an important secondary source of technology metals (Ongondo et al., 2015). However, their inclusion back in the economy has several bottlenecks, including technological limitations, and low collection rates of the devices and poor enforcement of the law concerning their management. These obstacles are interconnected, and amendment of one positively affects the other. In a circular economy, material loops are closed by recycling of discarded products, urban mining of EoL products and mining of current and future urban waste streams (Jones et al., 2013).

In this PhD research, novel metal recovery technologies from WEEE were investigated. An emphasis was given to biological methods. Biohydrometallurgy or urban biomining, using microbes for processing metals, enables environmentally sound and cost-effective processes to recover metals from waste materials (Ilyas and Lee, 2014a). In this context, microbial leaching (bioleaching) of metals from waste materials is an attractive field of research with vast potential. Moreover, conventional chemical technologies, despite their several bottlenecks and disadvantages, are effective in the leaching of metals from primary ores. However, their effectiveness in polymetallic, anthropogenic WEEE is largely unexplored. This research addresses the knowledge gap on two metal extraction approaches, namely chemical and biological, from a recent secondary source of metals, the essential parameters of these metal recovery processes, subsequent selective recovery techniques, techno-economic and sustainability assessment, and scale up potential of the technology.

1.2 Research goals and questions

The main objective of this work is *to develop a sustainable method to recover metals from electronic waste*. Moreover, the optimum process parameters are studied, different routes, e.g. biological and chemical, are explored and compared, as well an overall techno-economic and sustainability assessment of the newly developed technology were given. Application of biological methods in production of metals from primary sources is an established technology: more than 15% of Cu production is carried out by bacteria (Schlesinger et al., 2011). These biological processes are typically environmentally friendly, and cost effective processes, where the pollutant production is minimal and process input are simply the nutrient requirements.

The following hypotheses are formulated and tested:

1. **What is the best effective method to recover metals selectively?**

Recovery of metals from WEE is a necessity, in order to meet the demand for raw materials (Gu et al., 2016). Currently, there are several alternatives, such as pyrometallurgical, hydrometallurgical routes and recently emerging bio-based route, a technique that employs microbial cells to extract and recover metals from waste. Pyrometallurgy is an advanced refining technology, currently employed at full scale in commercial plants (Akcil et al., 2015). In this research, the most effective method is investigated and benchmarked to best available technologies (BAT).

2. **Which metals should be given priority to be recovered?**

Metals in WEEE are of variable abundance, chemical composition and form. They include base metals, precious metals and specialty metals (Reck and Graedel, 2012). The concentration and occurrence of individual metals depend on the type of waste, manufacture years and the source (Marques, 2013). There is no one-sizes-fits-all strategy to recover metals from electronic waste. Thus, it is essential to develop a waste- and metal-specific technology to recover metals from electronic waste. A number of relevant selection criteria include economic value of the metals, the criticality and technological barriers. This research question addresses the prioritizing of metal recovery from WEEE.

3. How effective are biological and chemical approaches?

Biological methods, leaching of metals mediated by microbial cells (bioleaching), are proven feasible for primary ores. Over 20% of Cu is produced by bacteria (Schlesinger et al., 2011). However, its application to secondary ores is largely unknown. Chemical processing, on the other hand, is an established technology for the primary ores and a number of approaches have been taken to suggested them from secondary ores, including WEEE (Cui and Zhang, 2008). However, due to dissimilar chemical composition of the materials to natural ores and large variety of metals, the recovery processes are fundamentally different. This research question investigates the effectiveness of two approaches, namely chemical and biological routes, in terms of selectivity, efficiency and scale-up potential.

4. How sustainable and feasible are the selected methods to recover metals from electronic waste?

Emerging bio-based technologies are regarded as environmentally friendly, as they do not emit hazardous gases or exploit corrosive reagents, and often autonomously remediate their potential hazards (Watling, 2015). Moreover, their energy requirement is minimal, as microbial cells require low energy to maintain their metabolism. This gives the biological route an advantage over conventional technologies. In this research, the sustainability assessment and techno-economic assessment of the newly developed technology are investigated.

1.3 Research approach and methodology

Complexity of the waste material requires innovative approaches in order to sustainably and selectively recover metals from WEEE. Therefore, a multidisciplinary approach is aimed, entrenching fields of environmental engineering, metallurgical engineering, environmental biotechnology, process development and sindustrial ecology. In this dissertation, the main dimensions were included for research goal achievement, namely, technology, sustainability, and the environment. For instance, components of leaching of metals from WEEE and its subsequent recovery are fundamentally different. Similarly, methods to develop a process, carry out cost analysis, and evaluate the environmental profile of a future technology are dissimilar. Nonetheless, it is required that those interrelated elements are analyzed altogether. Thus, a multidisciplinary approach was taken due to the complexity of the material, and this allowed to come out with a holistic overview of the collective outcome of all dimensions of sustainability.

1.4 Structure of this dissertation

The structure of this dissertation closely follows the research approach as given above in 1.3. Figure 1-2 illustrates the interaction of the chapters. The contents of the individual chapters are as follows:

Chapter 1: Introduction presents an overarching view of the entire dissertation, including a brief background of WEEE as a secondary source of metals and research methodology. The research goals and questions, the research approach and methodology, the scope and boundaries, the targeted audience are also given in this introductory chapter.

Chapter 2: Literature review gives a comprehensive literature analysis of the current status of WEEE and state of research on metal recovery. A holistic approach is taken and all elements of WEEE management are investigated. In this chapter, research gaps are identified after a critical analysis of the published literature.

Chapter 3: Literature review investigates the state-of-the-art on emerging biorecovery methods of metals from WEEE including novel bioleaching, bioprecipitation and biosorption technologies. In this chapter, perspectives on the knowledge gap on bioprocessing of a valuable secondary source is identified in the light of the recent developments in this field.

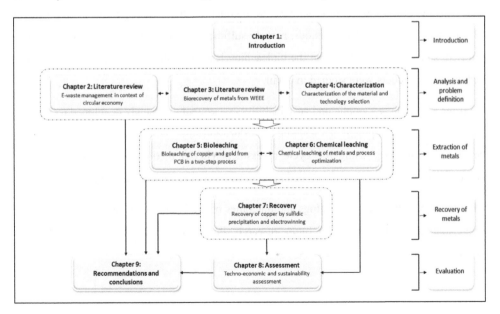

Figure 1-2: Structure of this PhD dissertation.

Chapter 4: Characterization and technology selection givens a comprehensive description of the discarded PCB materials from various devices, namely desktop computers, laptop computers, mobile phones and telecommunication devices. Also, technology selection using multi-criteria analysis (MCA) is given in this chapter.

Chapter 5: Bioleaching of metals presents the findings of multi-step bioleaching of Cu and Au from discarded PCB. This chapter demonstrates and discusses the experimental data from experiments of leaching Cu and Au in a two-step process by using *Acidithiobacillus ferrivorans* and *Acidithiobacillus thiooxidans*, followed by *Pseudomonas putida* and *Pseudomonas fluorescens* in the second step.

Chapter 6: Chemical leaching of metals discussed the results of an optimization study for the chemical leaching of metals from electronic waste using a two-step concept for Cu and Au. In this work, design of experiments (DoE) were conducted by central composite design methodology (CCD) for the two-step leaching of Cu and Au from electronic waste by using a mixture of sulfuric acid (H_2SO_4) and hydrogen peroxide (H_2O_2) in the first step, followed by mixture of thiosulfate ($S_2O_3^{2-}$) and ammonium hydroxide (NH_4OH) catalyzed by copper sulfate ($CuSO_4$).

Chapter 7: Selective recovery of metals from bioleaching solution studies the selective recovery of Cu from the real bioleaching solution obtained from the experiments performed in Chapter 5. In this work, two methods, namely sulfidic precipitation and electrowinning were selected, so as to recover Cu from the bioleaching leachate solution. The process parameters and the final products were studied in detail.

Chapter 8: Techno-economic and sustainability assessment of a newly developed technology focusses on the development of a scale-up procedure and sustainability assessment of the newly developed technology using life cycle assessment (LCA) methodology. Three alternative process route scenarios were developed in order to assess the various options.

Chapter 9: Summarizes the findings and outcomes of the PhD research and lists a number of suggestions and perspectives for future research. In Chapter 9, the finding of the PhD are discussed and the research outcomes

Chapter 2.

Electronic waste as a secondary source of metals, its management and recovery technologies

Abstract

The wealth of the society depends on a number of metals, including base metals, precious metals and increasingly rare earth elements (REE), collectively termed technology metals. Numerous applications stimulated the use of the technology metals, and their supply is at stake, owing to the high demand and uneven geographical distribution of these metals. Their stable supply is crucial for the well-being of the society and the transition to a sustainable, circular economy. There is an increasing interest in secondary sources of these metals. This chapter reviews the state of electronic waste, its management and latest technological developments in metal recovery from various streams of electronic waste. An emphasis was given to printed circuit boards (PCB), hard disc drives (HDD) and displays with regard to their detailed material content. Recovery technologies such as physical, pyrometallurgical and hydrometallurgical methods were overviewed respective to their level of technological readiness level (TRL). Finally, perspectives on electronic waste as a secondary source of metals are given.

2.1 Introduction

2.1.1 Definition of WEEE

Electronic waste refers to discarded electrical and electronic equipment that are at the end of their economic life span and no longer be used by consumers. It is commonly shortened as e-waste, and referred to as Waste Electric and Electronic Equipment (WEEE). All WEEE are grouped into 10 primary categories, according to the WEEE Directive by the European Commission (2012/19/EU). These 10 major product groups are classified per product type and legislative relevant categories. They are broken down into 58 sub-categories, approximately 900 products where all devices are represented. They are also linked to 5 to 7 collection categories, which exists in actual WEEE practice (Wang et al., 2012a).

2.1.2 Global and regional WEEE generation

WEEE is the fastest-growing type of solid waste, occupying an increasing fraction of the global municipal waste (Kiddee et al., 2013). This is particularly prevalent in the developed economies, with saturated electrical and electronic equipment (EEE) markets, where WEEE makes up to 8% of the municipal waste (Robinson, 2009). Generation of WEEE has exponentially increased, is associated to rapid technological innovations combined with demand growth in the electronics sector. In addition, decreasing economic lifespan of electronic devices (Zhang et al., 2012a), lack of international consensus on WEEE management (Friege, 2012), and inadequate user awareness play a role in an unprecedented increase of WEEE generation. The lifespan of electrical and electronic devices decreased from an average of 8 years to 2 years for large EEE and 4 years to 9 months for mobile phones, from 2000 to 2010, respectively (Kasper et al., 2011; Zhang et al., 2012a). Decreased economic use is particularly the case for urban areas where the population density is very high (Zeng et al., 2015). These issues, coupled with an ever-increasing spectrum of devices make the generation of WEEE an alarming issue.

The quantification of WEEE volumes is prerequisite for the development of sustainable solutions. Challenges include the lack of data accuracy and the dynamic behavior of the flows and their constituents (Schluep et al., 2013). WEEE quantification is particularly cumbersome in developing countries as informal waste management systems are poorly documented (Wang et al., 2013). Global WEEE generation reached 41.8 million tons (Mt) in 2014, and forecasted to rise to 50 million tons in 2018 (Baldé et al., 2015). An overview of EEE put on market, WEEE generation, and their projections until 2020 per country are given in Table 2-1 (Schluep et al., 2009; Ilyas and Lee, 2014a; StEP, 2015).

China plays a key role in the global electrical and electronic equipment (EEE) industry, in the manufacturing, the refurbishment, and reuse of EEE and recycling of WEEE. An increase in EEE usage and consequent WEEE generation are expected in China, and in other developing economies such as Brazil, Russia, India (BRIC group) and Turkey (Schluep et al., 2009). In the coming years, China will become the major WEEE producer, next its status as a primary EEE producer (Wang et al., 2013). There is a strong correlation between gross domestic product (GDP) and WEEE generation, and the economic development of a country is proportional with the amount of WEEE generated per person (Huisman, 2010). Thus, a sharp increase of WEEE generation in virtually all the developing countries is expected in the upcoming decades.

Table 2-1: Global growth of WEEE (Sources: Schluep et al., 2013; Ilyas and Lee, 2014b; StEP, 2015).

Country	EEE put on market in 2012 (Mt)	Annual estimated WEEE in 2013 (Mt)	Per inhabitant (kg/inh.)	Estimated in 2020	Increase
EU-27	9,800	10,205	19.6	11,430	12%
United States	9,350	9,359	29.3	10,050	7%
China	12,405	6,033	4.4	12,066	98%
Japan	3,300	3,022	23.8	3,200	5%
India	3,026	2,751	2.2	6,755	145%
Germany	1,752	1,696	21.9	1,974	16%
Russia	1,599	1,556	10.9	2,000	28%
Brazil	1,850	1,530	7.1	1,850	20%
France	1,520	1,224	21.6	1,625	32%
Italy	1,124	1,154	19.3	1,343	16%
Korea	959	961.3	19.2	1,050	9%
Turkey	726	661	8.8	800	21%
Netherlands	432	394	23.3	421	6%
Romania	217	157	7.9	227	44%
Norway	175	127	25.8	136	7%
Bulgaria	86	62	8.6	89	43%

2.2 Waste as a secondary resource in transition to a circular economy

The perspective in solid waste management has shifted from semi-engineered landfill disposal to recovery of materials and energy from secondary resources in the last decades. In a circular zero-waste economy, material flows are closed by the recycling of discarded products, and urban mining of current and future waste streams (Jones et al., 2013). Waste materials that are reprocessed to generate raw materials, potentially substituting the use of primary materials, are regarded as secondary resources (Ongondo et al., 2015). In this direction, a term 'urban mining' of EoL products is coined and widely accepted (2-1). The reintegration of wastes and by-products back in the economy strongly relies on the concept of waste as a secondary source of raw materials (Jones et al., 2013). In addition to being a hazardous waste, WEEE is an important secondary source of metals in transition to a circular economy and of primary importance in the context of urban mining of materials.

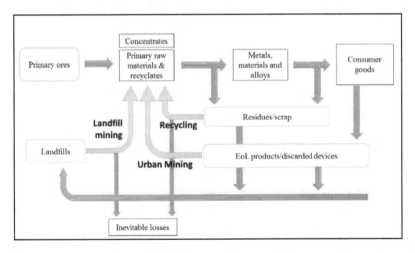

Figure 2-1: Closing material loops in a circular economy (Reproduced from Jones et al., 2013).

Several technical and non-technical tools have been developed to manage WEEE sustainably, taking a holistic approach that encompasses the entire chain of WEEE management, including collection, pretreatment, recovery and final disposal (Kiddee et al., 2013). Distinctive to primary mines, most secondary raw materials are rich in complex mixtures of metals, multi-element alloys, and polymetallic structures (Tuncuk et al., 2012; Ongondo et al., 2015). The complexity of the WEEE increased with the developing technology where more complicated alloys are invented and a great number of elements are included in electronic devices. Modern devices consist of up to 60 elements in various complex mixtures of metals (Bloodworth, 2014). From the recyclers point of view, highly complex alloys pose a challenge to develop an efficient metal recovery technology from WEEE.

2.3 Improper management of WEEE

2.3.1 Consequences of informal recycling

WEEE includes a large number of hazardous substances including lead-containing glass, brominated flame retardants (BFRs), polybrominated biphenyls (PBB), polychlororinated biphenyls (PCBs) and polybrominated diphenylethers (PBDEs) (Tsydenova and Bengtsson, 2011; Herat and Agamuthu, 2012; Zhang et al., 2012a). Environmental risks include the leaching of heavy metals (Lee and Pandey, 2012) and organic micro-pollutants (Oturan et al., 2015) into the groundwater from the landfill and stockpiles, and dioxin release from thermal processing of the waste material (Yang et al., 2013). In addition to the increasing volumes of WEEE generated, the list of WEEE-associated toxicants is also expanded (Zhang et al., 2012a) (Zhang et al., 2012a). Handling of WEEE in developing countries encompasses repair, reuse and substandard processing within a largely informal recycling sector (Ongondo et al., 2011). Solely in China, 690,000 people are estimated to be involved in informal collection and recycling (Robinson, 2009).

Typical informal substandard WEEE processing actions such as acid stripping and open burning, heating circuit boards for dismantling, melting of plastics release of hazardous chemicals contained in WEEE releases the toxic compounds as secondary pollutants (Tue et al., 2014). Substandard techniques are carried out in locations such as Guiyu in Guangdong Province in China, Taizhou in Zhejiang Province in China, Bangalore in India and Agbogbloshie in Ghana which are the major recipients of global WEEE (Breivik et al., 2014).

WEEE sites are 100 times more contaminated by PBDE, heavy metals and polycyclic aromatic hydrocarbons (PAHs) than other places (Ghosh et al., 2015). High levels of heavy metals in the vicinity of the recycling regions, elevated heavy metal content of the freshwaters and high dioxin concentration in the air validate the adverse effects of informal recycling to the local environmental quality (Hadi et al., 2015). Landfilling is the most common disposal method for WEEE, not only in developing countries but also in many developed countries, and approximately 40% of the WEEE goes to uncontrolled landfills (Cucchiella et al., 2015).

Handling and sustainable recycling of WEEE require novel strategies and technologies, owing to its peculiar composition. Unlike other types of hazardous waste, incineration of WEEE is dangerous (Wang and Xu, 2014). There are concerns on the toxicity of the gases produced, e.g. dioxins, furans, polybrominated organic pollutants, and polycyclic aromatic hydrocarbons (PAHs), which are the end products of thermal treatment of WEEE (Akcil et al., 2015). Moreover, the processes emit toxic heavy metal vapors and dioxins. Toxic polybrominated dibenzo-p-dioxins and dibenzofurans (PBDD/Fs) can be generated from incineration, and thermal treatment at 300°C or even by natural light (Duan et al., 2011).

In addition to all the hazards originating from WEEE, manufacturing mobile phones and personal computers consumes considerable fractions of the gold (Au), silver (Ag) and palladium (Pd) mined annually worldwide (Hadi et al., 2015). Electronics industry is the third largest consumer of Au, about 282 tons, accounting for 12% of the total Au demand in 2014 (Schipper et al., 2015). Worldwide, more than one million people in 26 countries across Africa, Asia and South America work in gold mining mostly in unregistered substandard conditions (Mccann and Wittmann, 2015), driven by the demand of this precious metal for electronics.

2.3.2 WEEE regulations

The collection and treatment of WEEE is regulated by an increasing number of countries in the world. In the EU, the recast of the WEEE Directive (2012/19/EU) brought higher standards compared to the earlier directive (2002/96/EC) in terms of collection objectives. It aims to solve the problems related to improper management of WEEE, which alongside the Restriction of Hazardous Substances (RoHS) Directive (2002/95/EC) complements the measures on preventing landfill and incineration. Moreover, it introduced the "take-back system" conferring the responsibility of WEEE collection on producers. In the EU, at least 85% of the waste generated should be separately collected from municipal waste by 2020.

In the United States (US), WEEE has long been not regulated and the producers are not required to contribute to the environmental and social costs of the problem (Ladou and Lovegrove, 2008). As of 2015, approximately half of the federal states of the US, regulate the WEEE in a patchwork of heterogeneous policies. China's legislative regulation titled "Regulations on the Administration of Recycling and Treatment of Waste Electrical and Electronic Equipment", came into force on January 2011. It dictated mandatory recycling of WEEE, implementation of extended producer responsibility (EPR), established subsidies for formal recycling and obligated certification for second hand equipment, aiming to limit the extent of the informal sector (Li et al., 2015). In India, where there is no separate collection system, waste is treated as municipal waste and the relevant regulation named "Hazardous Material Laws and Rules" aims to address the country's serious WEEE problem. South Korea, Japan and Taiwan follow the similar ''take-back'' trend as in the EU, ensuring manufacturer responsibility by setting an aim of 75% recycling of their annual production. In Japan, selective fractions of the WEEE stream are regulated, and the extent of the regulation is limited. A few countries in south East Asia, such as Malaysia, the Philippines, and Singapore, require pre-shipment procedures for the trade of used electronics, accepting imports for reuse purposes (Wendell, 2011).

2.4 Transboundary movement of WEEE

A decade ago, the main WEEE traffic routes were towards Asia, and especially China; however, the introduction of a tighter legislation in China urged new destinations to be chosen, such as Ghana, Nigeria, South Africa, Vietnam, India and the Philippines (Li et al., 2013). Between 16 - 38% of the WEEE collected in the EU (Sthiannopkao and Wong, 2013) and 80% in the US (Sthiannopkao and Wong, 2013) are sent to developing countries, mostly in form of used or discarded devices, for substandard recycling practices. An overview of suspected and known routes of WEEE trade is given in Table 2-2.

Despite the deterring legislation, its imperfect enforcement gives space for so called waste crime. The legal treatment of waste is more expensive than illegal operations, particularly in the developed countries. Transboundary movement of WEEE is restricted by the Basel Convention ratified by most countries of the world. Within Europe, pollution-based environmental crime, including illegal trade and disposal of hazardous waste is monitored by Interpol (Baird et al., 2014). The source of the problem derives from technologically immature solutions, poor or non-existing enforcement of the law, and waste being a commodity in contrast with the asymmetry of the regulations. Vulnerabilities exist throughout the entire WEEE supply chain, including collection, consolidation, transport, and treatment (Baird et al., 2014). Theft, lack of required permits, smuggling, and false load declarations are usually reported.

Table 2-2: Global WEEE trade (Sources: Breivik et al., 2014; Fonti et al., 2015; StEP, 2015).

Country / Region	Domestic Production (M t)	Export (M t)	Import (M t)
Exporters			
USA	6.6 - 9.4	3.3 - 5.6	-
EU-28	9.8	1.9 - 3.9	-
Japan	3.1	0.62	-
Importers			
China	3.1	-	2.0 - 6.0
India	0.36	-	0.85 - 4.2
West Africa	0.5	-	1.5 - 3.5

When exported to the developing economies, the costs of WEEE treatment are externalized (Mccann and Wittmann, 2015). Several initiatives have been undertaken to solve the global WEEE problem. The United Nations University (UNU) based StEP initiative aims to decrease the amount of WEEE by repair and reuse both by the producers and the consumers. EU-based initiatives such as GoodElectronics, Electronic Watch and US-based platforms such as Green Electronics Council contribute to the improvement of working conditions in the global electronics supply chain, with a specific focus on large producers.

2.5 Metals in WEEE

From the recyclers point of view, WEEE is a complex mixture of different materials in various concentrations. Electric and electronic equipment EEE consist of a mixture of metals, including base metals, precious metals and increasingly REE; along with plastic and other materials (Marques, 2013). The number of metals used in devices increased over the years, as the fluorescent, conductivity and alloying properties of technology metals were uncovered (Tunsu et al., 2015). Consequently, their usage in numerous technological applications increased.

Modern devices encompass up to 60 elements, with an increase of complexity with various mixtures of compounds (Graedel, 2011).

These elements are used to manufacture microprocessors, printed circuit boards, displays, and permanent magnets usually in tiny quantities and often in complex alloys (Bloodworth, 2014). Many of them are listed as the critical raw materials, which include many metals found in several units of WEEE (European Commission, 2014a). The critical metals have high economic importance and a high risk associated with their supply and economic usage. Secure supply of REE is an essential issue for the emerging energy- and fuel-efficient technologies, where they play a critical role in the manufacture of electric cars, wind turbines, and photovoltaics (Ongondo et al., 2015). Today, in the tide of decoupling economic growth from hydrocarbon dependency, the circular economy is under the risk of shortage of technology metals.

2.5.1 The material composition of WEEE

WEEE consists of a diverse range of materials and their concentrations largely depend on the type, manufacturer and age of the equipment (Stenvall et al., 2013). From a material flow point of view, it is possible to group four categories of materials, namely **(1)** ferrous metals, **(2)** non-ferrous metals, **(3)** plastics, and **(4)** others. After processing, the overall output from a WEEE treatment plant is 38.1% ferrous metals, 16.5% non-ferrous metals, 26.5% plastic, and 18.9% others (Bigum et al., 2012). Most metals in WEEE are found either in their native metallic form or as alloys of multiple elements embedded in nonmetallic components (Tuncuk et al., 2012; Sun et al., 2015). An overview of many uses of metals in electrical devices is given in Table 2-3. Several units and components of WEEE are concentrated in certain elements (Cucchiella et al., 2015). This review discusses three WEEE components in detail namely displays PCB, and hard disc drives (HDD), which are important secondary sources of critical metals. An illustrative explanation of the occurrence of metals in these units of electronic waste is given in Figure 2-2.

2.5.1.1 *Displays*

In line with the rapid innovation in the EEE, the display technology has changed significantly in recent years. Liquid crystal displays (LCD) are increasingly used and replaced the cathode ray tube (CRT) technology in the last decade (Kalmykova et al., 2015). Compared to CRT monitors, smaller size and lower costs favored the popularity of LCD. LCD are made of approximately 85% glass and are used in various equipment such as televisions, PC monitors, laptops, tablet PC, mobile phones (Rotter et al., 2013). Light Emitting Diode (LED) screens are the upgraded product of LCD screens and are expected to replace LCD in the short term (Cucchiella et al., 2015).

Indium (In) is an essential element for display technology, particularly in LCD, owing to its semiconductor and optoelectronic properties (Zhang et al., 2015). Indium tin oxide (ITO) films acts as the electrode in LCD and accounts for more than 70% of In use worldwide (Dang et al., 2014). It is found in a concentration range between 100 - 400 ppm, depending on the origin of the panels. In addition to In, LCD panels include (%, *w/w*) 3-5 Al, 0.5 - 3.1 Fe; (ppm) 260 Sn, 53 Mn, and 14 Mo (Rocchetti et al., 2015).

As an unevenly distributed scarce resource, In is included in the list of critical raw materials (European Commission, 2014a). Global In reserves are estimated to be 16,000 tones, a crustal concentration of approximately one sixth of Au. Its average concentration in primary ores is between 10 - 20 ppm, and is mainly produced as a by-product of Zn and Pb mining, often found in sphalerite and chalcopyrite minerals (Zhang et al., 2015). China holds a monopolistic market share of 73% of its production. In a business-as-usual scenario of the current In consumption rate, the reserves will deplete by 2025 (Hester and Harrison, 2009).

In recovery from discarded LCD is currently immature, and there is not yet a commercial application. The yield of In recovery from EoL LCDs is less than 1% (Buchert et al., 2012). The process of In mobilization is generally carried out by means of strong acids under different reaction conditions (Rocchetti et al., 2015). The main challenge in In recovery is the separation of ITO glass from LCD panels. Currently, there is a lack of automated commercial-scale processes to recycle high-volume LCD safely and economically (Zhang et al., 2015). This metals critical status is expected to increase due to its increasing demand and uneven global distribution.

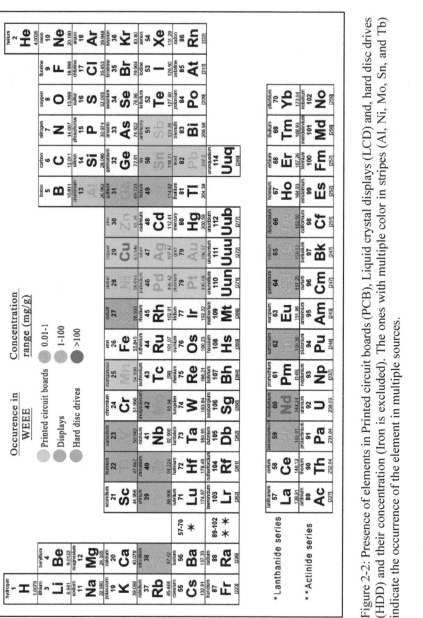

Figure 2-2: Presence of elements in Printed circuit boards (PCB), Liquid crystal displays (LCD), and hard disc drives (HDD) and their concentration (Iron is excluded). The ones with multiple color in stripes (Al, Ni, Mo, Sn, and Tb) indicate the occurrence of the element in multiple sources.

2.5.1.2 *Printed circuit boards (PCB)*

PCB support and connect the electronic components via conductive tracks, on etched copper sheets laminated onto a non-conductive glass-fiber reinforced epoxy resins substrate (Hall and Williams, 2007). They are a complex mixture of plastics, silicates, non-ferrous and ferrous metals, with variable concentrations. An overview of metal concentration of PCB is given in Table 2-4. All EEE, and most of their components, contain PCB differing in design and complexity (Cucchiella et al., 2015).

Bare PCB substrates represent about 23% of the total board weight (Duan et al., 2011). The weight fraction varies from 2% for large electronic devices (Dalrymple et al., 2007) to 11% for laptop computers (Fan et al., 2013) and up to 22% for mobile phones (Chancerel et al., 2009) of the total device weight. It is typically the major fraction of the total mass of a notebook, weighing around 250 - 300 g, coming third right after the LCD displays, and the battery (Fan et al., 2013). Many existing customized PCB designs makes it cumbersome to standardize a general PCB profile.

PCB are classified per their number of layers: **(a)** single layer, **(b)** double layer board in which circuits with two layers of copper are linked by metallized holes, and **(c)** multilayered boards in which the circuits are in multiple layers on a single board. This is crucial in defining its material composition, in particular the Cu content, the dominant material in PCB (Rubin et al., 2014). Metal concentrations of PCB depend on their source, type of the board, manufacturer, and period of production. Ever-changing manufacturing techniques, device-specific board designs and soldering technologies are the main drivers of diversity in discarded PCB (Marques, 2013). An illustration of a PCB and its components is given in Figure 2-3.

Table 2-3: Metals used in electrical and electronic equipment (EEE), their crustal and concentration, primary production, concentration in discarded devices, abundance and recycling rate.

Base metals

Metal	Form	Usage in EEE	Crustal conc. (ppm)	Concentration in ores (%, *w/w*)	World production from primary ores (kTons)	Demand for electronics (%)	Recycling rate (%)	Conc. in WEEE (%)	References
Cu	Elemental/ alloy	PCB, alloys, wiring, connectors, transformers	60	0.5 - 3	15,000	30	31	12 - 35	Graedel et al., 2011a; Yamasue et al., 2007; Ongondo et al., 2011; Hadi et al., 2015
Fe	Alloy	Steel, casing, magnets, casings	58,000	30 - 60	1,100,000	3	28	5 - 11	Graedel et al., 2011a; Yamasue et al., 2007; Hadi et al., 2015
Zn	Alloy/ oxide	Steel, Zn-Al-Cu alloy (94%), plating	80	0.15 - 0.65	13,000	12	18	0.1 - 2.5	Graedel et al., 2011a; Yamasue et al., 2007; Hadi et al., 2015;
Cr	Alloy	Steel (18%)	100	0.1 - 0.5	7,900	25	20	0.1 - 2.9	Graedel et al., 2011a; Yamasue et al., 2007; Hadi et al., 2015;
Ni	Elemental/ alloy	Steel (8%)	80	1 - 5	2,100	5	29	1 - 7.2	Graedel et al., 2011a; Yamasue et al., 2007; Hadi et al., 2015

Post-transition metals

Metal	Form	Usage in EEE	Crustal concentration (ppm)	Concentration in ores (%, w/w)	World production from primary ores (kTons)	Demand for electronics (%)	Recycling rate (%)	Concentration in WEEE (%)	References
Al	Elemental/alloy	Alloys, wiring, casing, heat sink	80,000	20 - 24	44,900	14	34	1.5 - 5	Graedel et al., 2011a; Yamasue et al., 2007; Hadi et al., 2015
In	Alloy/oxide	LCDs, semiconductors	0.5	0.1 - 0.2	0.6	79	<1	0.05 - 1	Graedel et al., 2011a; Zhang et al., 2015
Pb	Alloy	CRT funnel (14.7%), CRT neck (14.7%), solder (40%)	15	3 - 15	5,200	No data	63	0.9 - 5	Graedel et al., 2011a; Yamasue et al., 2007
Sn	Alloy	CRT, PCB	2.2	0.5 - 3	275	33	22	0.3 - 3	Graedel et al., 2011a; Herat and Agamuthu, 2012

Metalloids

Metal	Form	Usage in EEE	Crustal concentration (ppm)	Concentration in ores (%, w/w)	World production from primary ores (kTons)	Demand for electronics (%)	Recycling rate (%)	Concentration in WEEE (%)	References
Bi	Alloy	Solders, capacitor, heat sink	0.05	0.5 - 1	7.4	16	<1	0.05 - 3.5	Graedel et al., 2011a; Hadi et al., 2015;
Sb	Alloy	Flame retardant, CRT glass	0.2	0.1 - 10	180	50	20	0.1 - 0.7	Graedel et al., 2011a; Yamasue et al., 2007; Ongondo et al., 2011,

Metal	Form	Usage in EEE	Crustal concentration (ppm)	Concentration in ores (%, w/w)	World production from primary ores (kTons)	Demand for electronics (%)	Recycling rate (%)	Concentration in WEEE (%)	References
Sn	Alloy	CRT, PCB	2.2	0.5 - 3	275	33	22	0.3 - 3	Graedel et al., 2011a; Yamasue et al., 2007; Ongondo et al., 2011
In	Alloy/oxide	LCDs, semiconductors	0.5	0.1 - 0.2	0.6	79	<1	0.05 - 1	Graedel et al., 2011a; Herat and Agamuthu, 2012

Precious metals

Metal	Form	Usage in EEE	Crustal concentration (ppm)	Concentration in ores (%, w/w)	World production from primary ores (kTons)	Demand for electronics (%)	Recycling rate (%)	Concentration in WEEE (%)	References
Au	Alloy/elemental	PCB, contacts, integrated circuits (ICs)	0.002	5 - 10	2.35	12	43	30 - 350	Graedel et al., 2011a; Ongondo et al., 2011
Ag	Alloy	PCB, Brazing alloy (3%), lead-free solder (3%)	0.08	5 - 10	20	30	16	80 - 1000	Graedel et al., 2011a; Yamasue et al., 2007; Ongondo et al., 2011
Pd	Alloy/elemental	MLCC, PCB	0.0005	1 - 10	0.3	14	50	30 - 200	Graedel et al., 2011a; Herat and Agamuthu, 2012

Lanthanides

Metal	Form	Usage in EEE	Crustal concentration (ppm)	Concentration in ores (%, w/w)	World production from primary ores (kTons)	Demand for electronics (%)	Recycling rate (%)	Concentration in WEEE (%)	References
La	Alloy/Oxide	Lenses, batteries, alloys	18	0.5 - 20	32	No data	<1%	-	Binnemans et al., 2013; Simoni et al., 2015

Metal	Form	Usage in EEE	Crustal concentration (ppm)	Concentration in ores (%, *w/w*)	World production from primary ores (kTons)	Demand for electronics (%)	Recycling rate (%)	Concentration in WEEE (%)	References
Dy	Alloy/Oxide	Permanent magnets, HDD	0.3	0.1 - 8.6	0.1	No data	<1%	1.4	Cotton, 2006, Ueberschaar and Rotter, 2015; Binnemans et al., 2013;
Nd	Alloy/Oxide	Permanent magnets, HDD	24	0.1 - 15	19	82	<1%	5.0-22.9	Cotton, 2006, Ueberschaar and Rotter, 2015; Binnemans et al., 2013
Pr	Alloy/Oxide	Permanent magnets, HDDs	5.5	0.1 - 4.5	6.1	No data	<1%	1.5 - 2.5	Ueberschaar and Rotter, 2015; Binnemans et al., 2013; Simoni et al., 2015
Y	Alloy/Oxide	Florescent phosphors, alloys, LCDs	n/a	0.05 - 2.5	No data	No data	<1%	5 - 15.5	Cotton, 2006; Binnemans et al., 2013; Innocenzi et al., 2014; Simoni et al., 2015

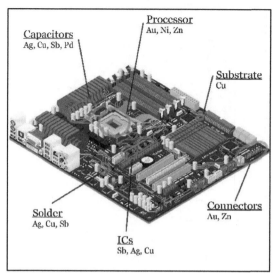

Figure 2-3: Elements found in printed circuit boards (PCB) shown in individual parts of the unit.

Cu, the predominant metal in PCB, forms the conducting layer for electrical connection. The Cu concentration of PCB positively correlates with the number of its layers and the capacity of conductor current (Hadi et al., 2015). Metals are accumulated on certain components of PCB, i.e. microprocessors, chips, connectors, and capacitors. The Cu concentrations of PCB are between 15-35% by weight. Platinum group metals (PGM) are used as contact materials in joints relays, switches or in sensors due to their high chemical stability and their good conducting properties. Palladium (Pd) is used for multi-layer ceramic chip capacitors (MLCC) and microprocessors (CPU). Gold (Au) is used in contacts, soldered joints, connecting wires and microprocessors, as contact metal in a thin film, often alloyed with small amounts of nickel (Ni) or cobalt (Co) to increase durability (Luda, 2011). CPU are the components of PCB on which Au is mainly accumulated, these units may contain up to 1900 ppm Au (Birloaga et al., 2013). Au content reasonably decreased in years from a contact layer of $1-2.5$ µm in the 1980s to 300 - 600 nm thick layer in modern appliances (Widmer and Oswald-Krapf, 2005; Cui and Zhang, 2008). Ni in its pure form is also increasingly used in contacts as additive as well as finishing. Lead (Pb) in solders has been replaced following the restriction of its use in electronics by the Restriction of Hazardous Substances in Electrical and Electronic Equipment (RoHS) directive (2011/65/EC). Integrated circuits and semiconductors consist of small amounts of gallium (Ga), indium (In), titanium (Ti), germanium (Ge), arsenic (As), selenium (Se) and tellurium (Te). Heat sinks are predominantly beryllium (Be) in form of beryllium oxide and Al, two compounds used for their heat conductivity. Sn is also located on the surface of PCB (Marques, 2013). Chromium (Cr) and (Zn) are used as additives to steel (Ghosh et al., 2015).

2.5.1.3 *Hard disk drives*

Hard disk drives (HDD) are an important secondary source of rare earth elements (REE), particularly neodymium (Nd), praseodymium (Pr), and dysprosium (Dy). These elements are used for their magnetic properties in various types of complex alloys employed in permanent magnets. The most widely used alloy is a mixture of Nd, Fe and boron (B) with the acronym NdFeB. The concentrations of Nd, Pr and Dy in permanent magnets are (in %, by weight) 22.9, 2.7 and 1.4, respectively (Ueberschaar and Rotter, 2015). In 2008, six hundred million HDD were manufactured, each containing approximately 20 g of NdFeB (Tunsu et al., 2015).

Typically, a desktop PC HDD weights around 542 (± 68) g, a share of 3.3% of an average PC with a weight of 12.2 kg. Unlike PCB, the weight fraction of HDD of PC and laptops are comparable, as a HDD occupies 3.6% of the total weight of the laptop device, in average 3.2 kg HDD is 134 (± 26) g (Ueberschaar and Rotter, 2015). Storage technologies undergo an upgrade and HDD phase out and are replaced with by Solid State Disks (SSD) (Cucchiella et al., 2015). Their substitution is, however, slower than compared to the transformation of LCD to CRT. Recovery of critical metals from HDD is very low due to thermodynamic limitations, very dense and complex structures of the permanent magnets, and lack of liberation and separation technologies. Separating the permanent magnets is an uneasy task due to compact product design, the use of a wide variety of screws and the magnets being strongly attached to other components (Tunsu et al., 2015). This applies especially for REE that are present in complex alloys and the NdFeB magnets that are difficult to access to.

2.6 Metal recovery from WEEE

Significant efforts are made to develop product- and metal-specific recovery processes from WEEE. Use of secondary sources allow the conservation of primary ores, thus significantly reduce the carbon and ecological footprints (Tuncuk et al., 2012). Additionally, up to 95% and 85% energy is saved when recycled, respectively, for A) and Cu compared to production from primary ores (Cui and Forssberg, 2003). Currently, 50% of the semi-finished Cu products come from recycled materials (Schlesinger et al., 2011).

Most of the industrial metal recovery processes from WEEE are restricted to physical pretreatment and pyrometallurgical processes (Cui and Zhang, 2008). Current best available technologies (BAT) are imperfect, not metal-selective, environmentally malign and energy intensive (Mäkinen et al., 2015). Effective technologies are lacking due to the complex nature of the waste material, thus requiring eco-innovative urban mining solutions. Various metal recovery approaches are currently in operation or research and development phase, including pyrometallurgy, hydrometallurgy and biohydrometallurgy.

Pyrometallurgical processes, encompassing smelting and pyrolysis, require the heating of WEEE at very high temperatures to separate materials. In industrial processes, WEEE is fed immediately after physical separation. Hydrometallurgical treatments are based on the use of leaching agents in aqueous solutions, such as strong acids and bases, often applied together with other complexing agents. Biohydrometallurgy is based on similar principles where the lixiviants are biologically produced. The rates of hydrometallurgical processes are relatively faster than those of biohydrometallurgical processes, whereas biological processes are more environmentally friendly and cost-effective (Ilyas and Lee, 2015). Moreover, bioleaching is selective towards individual metals as microorganisms are sensitive to high concentrations several metal ions. Lastly, biological processes have vast potential to improve as the microbes adapt to toxic conditions and eventually increase their metal tolerance capacity (Navarro et al., 2013). Recently, hybrid technologies have also been applied, which integrate the chemical processes (more efficient) with biological processes (more environmentally friendly), thus taking advantage of the benefits of both chemical and biological leaching (Pant et al., 2012; Ilyas et al., 2015).

Table 2-4: Metal content of various PCB.

PCB type	Pre-treatment	Particle size (mm)	Dissolution method	Analytical instrument	Weight (%)					Weight (mg/kg)			References
					Cu	Fe	Al	Pb	Ni	Au	Ag	Pd	
PC	Magnetic separation	0.25 - 0.5	A.R.	AAS*	23.5	0.1	1.6	1.0	0.2	n.m.	n.m.	n.m.	Veit et al., 2006
PC	Manual dismantling	Not specified	Fire assay	ICP†	20	7	5	1.5	1	250	1000	110	Hagelüken, 2006
PC	Mechanical + magnetic separation	Various <2 mm	A.R. at room temperature for 24 h	ICP-OES‡	20.2	7.3	5.7	5.5	0.4	0.21	0.16	n.m.	Yamane et al., 2011
PC	Manual dismantling + magnetic separation	<0.5	A.R. at room temperature for 18 h	ICP-MS§	17.3	2.1	2.1	12.5	7.2	0.210	n.m.**	n.m.	Janyasuthiwong et al., 2015
Mobile	Manual dismantling	Not specified	Fire assay	ICP	13	5	1	0.3	0.1	350	1340	210	Hagelüken, 2006
Mobile	Mechanical + magnetic separation	Various <4 mm	A.R. for 24 h	ICP-OES	34.5	10.6	0.3	1.9	2.6	n.d	210	n.m.	Yamane et al., 2011

* A.R.: Aqua regia: Nitro-hydrochloric acid (a mixture of hydrochloric acid and nitric acid in a ratio of 3 to 1, by volume, respectively.
† ICP: Inductively coupled plasma
‡ ICP-OES: Inductively coupled plasma – optical emission spectrometry
§ ICP-OES: Inductively coupled plasma – optical emission spectrometry
** Not measured

PCB type	Pre-treatment	Particle size (mm)	Dissolution method	Analytical instrument	Weight (%)					Weight (mg/kg)			Reference
					Cu	Fe	Al	Pb	Ni	Au	Ag	Pd	
Mobile	Manual dismantling + magnetic/electrostatic separation	<1 mm	A.R. at 60 °C for 2 h	ICP-AES	39.6	1.4	0.3	1.2	3.4	600	600	n.m.	Kasper et al., 2011
TV†† (CRT removed)	Magnetic separation	ground	A.R.	ICP-AES, ICP-MS	3.4	n.m.	1.2	0.2	0.04	<10	20	<10	Cui and Forssberg, 2007
PCB (not specified)	Washed in NaCl solution	100-120 µm	A.R.	AAS	8.9	8	0.75	3.15	2.0	13	30	n.m.	Ilyas et al., 2010
Mixed	Mechanical processing, Fe and Al removed	ground	A.R.	Not specified	28.7	0.6	1.7	1.3	n.m.	79	68	33	Creamer et al., 2006
Laptop	Manual dismantling	<0.5	A.R	ICP-MS	17.61	3.74	1.9	2.24	5.73	0.305	n.m.	n.m.	Chapter 4
Printer	Crushing, milling	<0.6	A.R	AAS, ICP-AES	19.2	3.56	7.06	1.0	5.4	70	100	n.m.	Yoo et al., 2009

†† TV: Television

2.7 Physical pretreatment of WEEE

Mechanical processes aim to liberate and separate the encapsulated metallic elements and enable their extraction. It is usually the first step in resource recovery from waste materials. Several physical treatment methods have been developed, including:

- manual and semi-automatic routes
- comminution
- gravity separators
- magnetic sorting
- optical separation

Physical treatment also includes the removal of hazardous materials, such as Hg from the backlight of LCD (Zhang et al., 2015) and toxic Be from PCB (Wang et al., 2013). The high complexity of these components necessitates a meticulous approach to liberate the metals of interest from separate components. However, loss and contamination of critical metals such as platinum group metals (PGM) (Chancerel et al., 2009), and REE (Ueberschaar and Rotter, 2015) have been reported with the existing mechanical processes. Moreover, physical crushing units are not yet suitable for the processing of such waste streams and contamination is inevitable (Yoo et al., 2009; Wienold et al., 2011).

2.7.1 Manual sorting and separation

Many metals are concentrated on certain parts of the WEEE components, thus a manual separation is essential. In a recycling process, disassembly of these parts is the most time-consuming operation. Durations of the disassembly and separation of the main case, the PCB and the LCD panel lasts 133, 67 and 64 s, respectively (Fan et al., 2013). Discarded LCD need to be dismantled to break the plastic shell, and remove the indium trioxide (ITO) panels (Zhang et al., 2015). Automatic, semiautomatic and manual disassembly systems have been developed, and the latter is the commonly adopted technique (Cimpan et al., 2015). The recovery efficiency by manual treatment is a lot higher than that of automatic systems (Zhang et al., 2015). On the other hand, from an economic point of view, manual sorting and dismantling is not feasible to be applied in developed economies, where labor costs are very high. To this end, the so-called ''Best-of-two-Worlds'' (Bo2W) approach has been proposed (Wang et al., 2012b). The Bo2W philosophy argues the combination of the developing economies' experience in manual dismantling with the developed economies' expertise in metal processing for the most efficient recovery technologies.

2.7.2 Size reduction/comminution

Manual sorting and dismantling is typically followed by a size reduction step. Various types of crushers, shredders, cutters equipped with a bottom sieve are used for liberation, for almost all types and components of WEEE (Yoo et al., 2009). Conventional crushers operate poorly due to the presence of reinforced resin, copper layer and glass fiber. In contrast, shredding or cutting, which works on the principle of shearing, is more effective (Wienold et al., 2011).

After comminution, most of the particles (25%) are accumulated in the 1 - 2 mm particle size, followed by the 0.5-1 mm (18%) and the 2-3 mm (17%) particle sizes, respectively. The PSD of particulate materials can be generalized using the Rosin–Rammler distribution function as expressed in:

$$(R(x) = 100e^{-(Ax)^B} \tag{2-1}$$

where R(x) is the cumulative oversize mass, (wt.%), x is the particle size, (in mm), 1/A is the Rosin–Rammler geometric mean diameter (mm); and B is the Rosin–Rammler skewness distribution parameter.

Janyasuthiwong et al. (2015) found that the Cu concentration of the particle size fraction 0.5 - 1.0 mm was higher than that of particle size <0.5 mm. This observation can be explained in terms of the different comminution mechanisms of the mills. When a brittle material such as PCB is milled, the up-and-down reciprocating motion of the stamp mill hammer concentrates the finer particles. Similar to PCB, size-reduction is the applied method to separate for the LCD by removing the polarizing film and liquid crystal (Zhang et al., 2015).

Physical separation is a common technique to process all types of WEEE. However, the one size-fits-all approach proves inefficient for such a complex type of waste. Moreover, high energy consumption, relatively low efficiency, as well as loss and contamination by metals are important obstacles in physical processing of WEEE for metal recovery (Chancerel et al., 2009; Zhang et al., 2015). Thermodynamic limitations necessitate novel liberation and separation strategies, particularly for the metals embedded in non-metal components (Ueberschaar and Rotter, 2015).

2.7.3 **Corona-electrostatic and Eddy-Current separation**

The diverse material composition of WEEE enables the separation of materials based on their electrical conductivity difference. In a corona separator, particles pass over a high-speed drum equipped with high-energy permanent magnet drum to separate plastic particles, non-ferrous metals and ferrous metals (Li et al., 2014b). Eddy-current based electrostatic separators operate based on similar principle of the conductivity difference (Jujun et al., 2014). The small particles, typically less than 0.6 mm, are passed along a vibratory feeder to a rotating roll to which a high voltage electrostatic field is applied using a corona and an electrostatic electrode (Bigum et al., 2012). The separation efficiency depends on the different trajectories of particle movements due to eddy current, induced in the non-ferrous particles, and the external magnetic field which deflects the particles respective to their conductivity (Ghosh et al., 2015).

2.7.4 **Magnetic separation**

While separating metals according to their magnetic properties, Fe and Ni are typically accumulated in the magnetic fraction, and Cu in the conductive fraction (Kasper et al., 2011). As similar with electrostatic separation, this process is typically applied after comminution. Thus, the particle size plays an important role. A minor fraction could be left in the non-magnetic fraction, probably due to their presence in paramagnetic or diamagnetic particles as alloying elements (Yoo et al., 2009). The magnetic fraction of the crushed PCB is reported to vary between 4.5% and 11% of the total weight (Janyasuthiwong et al., 2016). Magnetic separators are inefficient for the crushed PCB (Ghosh et al., 2015) and LCD (Zhang et al., 2015) due to agglomeration and consequent loss of non-magnetic materials. This method is poorly selective towards metals and unable to separate individual metals from their alloys.

2.7.5 **Gravity separation**

Gravity separation depends on the density and the size of the particles: their movement in a fluid, e.g. air, allows the separation of the different particles. One of the major disadvantages of this method is the simultaneous difference of the particle size and density. When the particle size of the crushed material is smaller than 0.45 mm, the grade of the copper is drastically decreased (Li et al., 2014a). This is, however, due to the issues associated with particle size distribution (comminution), as discussed earlier (See section 2.7.2).

Yoo et al. (2009) applied gravity separation using a zig-zag classifier to separate the metallic and non-metallic components of ground PCB. The separation of the metallic components into the heavy fractions increased with increasing particle size. About 95% of the metallic components were separated into the heavy fractions from the milled printed circuit boards of a size smaller than 0.6 mm, but this was remarkably reduced to 60% as the particle size of the milled printed circuit boards became smaller than 0.6 mm.

2.8 Treatment and refining of WEEE

2.8.1.1 *Thermal treatment*

Smelting is currently the industrial best available technology (BAT) and several WEEE processing plants already operate in Europe. At Boliden Ltd. Ronnskar smelter in Sweden, discarded PCB are directly fed into a copper converter to recover Cu, Ag, Au, Pd, Ni, Se and Zn (Ghosh et al., 2015). At Umicore's integrated metal smelter and refinery in Hoboken, Belgium, PCB are first treated in an IsaSmelt furnace to recover precious metals. It is further refined with hydrometallurgical processes and electrowinning (Zhang and Xu, 2016).

The disadvantages of smelting are high-energy consumption, high environmental effects, and low selectivity towards individual metals. Smelters have long been considered as significant sources of hazardous SO_2 and toxic heavy metals (Cappuyns et al., 2006; Zhang et al., 2012b). Two tons of SO_2 are emitted in flue gases per ton of copper produced from primary ores (Carn et al., 2007). In addition, the formation of dioxins and other gaseous emissions due to existing flame retardant chemicals are prevalent (Mäkinen et al., 2015). Moreover, a large range of WEEE are not suitable to be directly processed by a smelting process because of their low calorific values (Sun et al., 2015).

2.8.1.2 *Pyrolysis and gasification*

Similar to conventional incineration, pyrolysis targets the organic fraction of WEEE. Pyrolysis of discarded PCB, carried out at elevated temperatures up to 700°C in the presence of inert gases, generates 23% oil, 5% gases and 70% metal-rich residue (Hall and Williams, 2007). As such, discarded LCD panels are subjected to pyrolysis in ceramic ovens at 700°C, and an organic-rich polarizing film was converted into pyrolysis oil and gas, and the liquid crystal is eliminated (Ma and Xu, 2013). However, this method is inefficient due to the high costs associated with high energy and reagent consumption. Moreover, pyrolysis is potentially a dangerous method, as similar to incineration, toxic compounds are formed at high temperatures. A certain amount of bromine is found in the char or ash product, possibly due to the brominated flame retardants (BFRs) content of the discarded PCB. The composition of the gaseous products of pyrolysis and gasification are identical (Havlik et al., 2010).

2.8.1.3 *Electrochemistry*

Direct and indirect strategies have been investigated for the recovery of metals from WEEE using electrochemical routes. The electrical current can be manipulated to oxidize and reduce metals in redox reactions, in order to mobilize them from the waste material. Kasper et al. (2011) studied the effect of the electrode type, current density and exposure time on the process efficiency. The ground mobile phone PCB were leached with nitro-hydrochloric acid and subjected to electrowinning at 6 A/dm^2 current density. 93% of the dissolved Cu (5 g/L) was electrodeposited after 90 min. Due to loss of adherence on the cathode, Kasper et al. (2011) recommended the increase of cathode size.

Pilone and Kelsall (2006) used electro-generated chlorine to leach metals, followed by electro-deposition of metals at a graphite felt cathode as counter reaction of anodic generation of chlorine. Fogarasi et al. (2014) made a comparative environmental assessment of the Cu-recovery from a WEEE leachate by two different electrochemical routes. Both electrochemical processes involved the dissolution of copper from waste PCB with simultaneous cathodic electrodeposition from the resulting leach liquor. The first process used direct electrochemical oxidation in sulfate medium, while in the second dissolution of Cu was through mediated electrochemical oxidation of the Fe^{3+}/Fe^{2+} redox couple in chloride medium. The latter process was found to have a lower environmental impact because the regenerated the Fe^{3+}/Fe^{2+} solution can be used for further processing without the addition of a fresh reagent.

2.8.2 Hydrometallurgical treatment of WEEE for metal recovery

Hydrometallurgical metal recovery processes involve an oxidative leaching for the extraction of metals, followed by separation and purification procedures (Schlesinger et al., 2011). It has advantages over pyrometallurgy such as lower toxic residues and emissions, and higher energy efficiency. However, these processes still pose a threat due to the use of large amounts of toxic, corrosive and flammable reagents and the generation of high volumes of effluents and other solid wastes (Tuncuk et al., 2012).

2.8.2.1 *Acid leaching*

Acid leaching of metals from WEEE has been investigated using various acids and oxidants, or mixtures thereof. It is an essential process during the extraction of valuable metals from PCB

(Ghosh et al., 2015), indium from ITO glass (Zhang et al., 2015), and Nd from HDD (Li et al., 2009). In oxidative acid leaching, the important parameters are temperature, concentration and contact time with the former being the most important. The leaching of metals in various oxidative acidic media have been investigated for their effectiveness in metal recovery from waste PCB, including HCl (Jha et al., 2012), H_2SO_4 (Kumar et al., 2012), HNO_3 (Joda and Rashchi, 2012), sodium hypochlorite (NaOCl) (Akcil et al., 2015), thiosulfate ($S_2O_3^{2-}$) (Ha et al., 2010), thiourea (Jing-ying et al., 2012), halides (Syed, 2012).

Indium (In) in discarded LCDs reacts with sulfuric and hydrochloric acid under elevated temperature conditions (Rocchetti et al., 2015). Al and Sr leaches in concentrated HCl, while HNO_3 and concentrated H_2SO_4 are more selective towards In. Kato et al. (2013) reached 90% In leaching by 3.2 M (10%, v/v) HCl. Ruan et al. (2012) used H_2SO_4 at a liquid to solid ratio (L/S) of 1:1 at 160°C for 1 h and reached 92% of In leaching. Wang et al. (2013) leached 100% In leaching with 0.6 M H_2SO_4 in 42 min, and at a temperature of 68.6°C. In leaching showed a positive correlation with temperature and acid concentration in all the experimental work this far reported. Time has little effect on the final In leaching efficiency.

2.8.2.2 *Cyanide leaching of precious metals*

Cyanide leaching is the industrial norm for the leaching of precious metals from their primary ores (Zhang et al., 2012c). It interacts with nearly all transition metals, except lanthanides and actinides, and forms complexes with high chemical stability (Marsden and House, 2006). Out of 875 Au and Ag mines being operational in 2000, more than 90 used cyanide as lixiviant (Akcil et al., 2015). Au is bound by cyanide in a reaction named the Elsner's equation:

$$4Au + 8CN^- + O_2 + 2H_2O \rightarrow 4Au(CN)_2^- + 4e^- \tag{2-2}$$

Cyanide is preferred because of its cost-effectiveness. However, effluent treatment is problematic as cyanide is very lethal to most life forms. Several non-cyanide leaching processes, have been developed considering toxicity issues and handling problems. However, none of them has yet proven to be more cost-effective than cyanide (Akcil, 2010).

2.8.2.3 *Thiosulfate leaching of gold and silver*

Thiosulfate leaching attracted interest as an alternative precious metal leaching agent owing to its environmental advantages (Zhang, 2008). Thiosulfate leaching can be considered as a non-toxic process, the Au dissolution rates can be faster than conventional cyanidation (Aylmore and Muir, 2001). In alkaline or near neutral solutions of thiosulfate, Au dissolves slowly in the presence of a mild oxidant, as given in Equation (2-3).

$$Au + 2S_2O_3^{2-} \rightarrow Au(S_2O_3)_2^{3-} + e^- \tag{2-3}$$

Several studies investigated the leaching of precious metals from waste PCB, such as the effect of the thiosulfate concentration, alkalinity agent (e.g. NH_4OH) and catalyzing agent (e.g. Cu ions) (Ha et al., 2010; Ficeriová et al., 2011; Petter et al., 2014). Similar to cyanide, dissolved Cu may adversely affect the leaching process through the decomposition of thiosulfate. The reaction is relatively inefficient and slow under ambient temperatures. High consumption of the leaching agent, its chemical instability and low cost-effectiveness make it less effective compared to cyanide leaching (Akcil et al., 2015).

2.8.2.4 *Thiourea leaching of gold*

Thiourea ($SC(NH_2)_2$) is an organosulfur compound that forms white crystal complexes with many transition metals. Under acidic conditions, with the presence of an oxidant such as Fe^{3+}, ($SC(NH2)2$) and Au will form soluble cationic complexes with Au (Li and Miller, 2007). The rate of Au dissolution is strongly determined by pH. The role of Fe^{3+} in the complexation process is to facilitate the oxidation of metallic gold (Au^0) to the aurous (Au^+) ions (Gurung et al., 2013). The dissolving reaction of Au in $SC(NH_2)_2$ with Fe^{3+} as the oxidizing agent occurs according to Equation (2-4).

$$Au + 2CS(NH_2)_2 + Fe^{3+} \rightarrow Au[CS(NH_2)_2]_2^+ + Fe^{2+} \tag{2-4}$$

Several strategies including supply of an additional oxidant or, two-step leaching procedure were proposed for the leaching of precious metals by $SC(NH_2)_2$ (Jing-ying et al., 2012; Behnamfard et al., 2013). Elevated temperatures proved to be inefficient due to poor thermal stability of the reagent (Gurung et al., 2013). The high cost and chemical instability of $SC(NH_2)_2$ are challenges for the development of a scaled-up process.

2.8.2.5 *Halide (iodine and chlorine) leaching of gold*

Iodine (I_2) and chlorine (Cl_2) can act as redox, complexing and precipitating agents under certain conditions. This property gives them an advantage to achieve selective recovery of PGM from waste materials (Serpe et al., 2015). The dissolving gold reaction with chloride and iodine are given below in Equations (2-5) and (2-6) respectively:

$$Au + 4Cl^- \rightarrow AuCl_4^- + 3e^- \qquad (2\text{-}5)$$

$$2Au + 3I_3^- \rightarrow 2AuI_4^- + I^- \qquad (2\text{-}6)$$

Several approaches including leaching of metals using electro generated chlorine (Kim et al., 2011b), Au leaching using an iodine- hydrogen peroxide (I_2-H_2O_2) system (Sahin et al., 2015), I^-/I_2 leaching of Au in a three-step leaching system (Serpe et al., 2015) have been investigated. Despite the fast kinetics and high efficiencies of this method, very high rate of reagent consumption and reagent costs are the main obstacles of this process.

2.8.3 Biohydrometallurgical treatment of WEEE

Biohydrometallurgy, the use of microbes to process metals, is an efficient technology for the production of metals from primary ores (further discussed in Chapter 3). It is an established technology to process many metals including Cu, Au, nickel Ni, Zn, cobalt (Co), , arsenic (As), molybdenum (Mo), cadmium (Cd), and uranium (U) (Watling, 2015). More than 15% of the total 15 M t of Cu and 5% of Au, along with small fraction of Ni and Zn are produced using biohydrometallurgical routes (Schlesinger et al., 2011; Johnson, 2014). Recent developments include the use of acidophilic bacteria, cyanogenic heterotrophs and acid-producing heterotrophs to selectively recover metals from waste streams. Full scale biomining applications compensate high initial capital investments with lower operating cost over a long period (Brierley and Brierley, 2013). Furthermore, biotechnology relies on natural material cycles in more environment friendly processes than conventional metal extraction techniques.

Biotechnological approaches will play a significant role in the treatment of wastes for metal recovery (Lee and Pandey, 2012). Despite the relative slower kinetics compared to conventional methods, bioleaching has matured into a well-developed technology operated in advanced engineered systems. Biomining of low grade ores, and in particular Cu, made a useful case study for bioprocessing of metals, as expected to be the future trend for non-sulfide metal-rich wastes such as WEEE (Orell et al., 2010).

As similar to primary ores, there is an increasing interest on WEEE bioprocessing for metal recovery, owing to its **(1)** better environmental profile compared to conventional methods, and **(2)** high potential for future improvement. Most reported research work is performed at low technology readiness level (TRL), with a few scale up applications. Recent findings, such as involvement of contact mechanism (Silva et al., 2015) and waste load up to 10% (*w/v*) (Ilyas and Lee, 2014b), are important developments in WEEE bioprocessing. Full-scale applications require further investigations, optimization of the operational conditions and evaluation of costs and ex ante environmental impacts. In addition, complementing scale-up studies with techno-economic assessment and environmental sustainability analysis considerations are the ways to proceed in WEEE bioprocessing research.

Reported Cu bioleaching efficiencies vary widely from 50% to 100% with leaching periods typically exceeding than 7 days and pulp densities of 1 - 3% (w/v). Several studies have demonstrated improved bioleaching efficiencies in sulfur- and ferrous iron-supplemented media (Ilyas et al., 2010; Liang et al., 2010; Wang et al., 2009). Metal removal efficiencies reportedly decrease significantly with increasing pulp density (Erüst et al., 2013). This has several possible causes. Firstly, the waste material has an alkaline nature, and is therefore acid-consuming (Brandl et al., 2001). This results in a high pH environment, where the acidophiles do not thrive. Secondly, the non-metallic fraction, i.e. epoxy-coated substrate, and organic fraction of the material can be toxic to the bacteria (Niu and Li, 2007; Zhu et al., 2011).

Recently, most research is focused on enhancing the bioleaching rate of the metals from WEEE and increase the waste load in order to acclimatize the bacteria and be competitive with chemical technologies. In a first attempt, Ilyas et al. (2007) reached a final leaching efficiency of 80%, 64%, 86% and 74% for Zn, Al, Cu and Ni, respectively after a pre-leaching period of 27 days followed by a bioleaching period of 280 days. They used a column setup for bioleaching of metals from WEEE by moderately thermophilic acidophilic bacterial strains of chemolithotrophic and heterotrophic consortia. In a follow up research, they reached 91%, 95%, 96% and 94% efficiency for Al, Cu, Zn, and Ni respectively, at a pulp density of 10% (w/v) in a continuous stirred tank reactor (CSTR) bioleaching setup using a moderately thermophilic adapted culture of *Sulfobacillus thermosulfidooxidans*. The bioleaching mixture was supplemented with enriched air of 25% O_2 + 0.03% CO_2, with 2.5% (w/v) biogenic S^0 and maintained at 45°C. An interesting finding was the faster oxidation rate of biogenic sulfur over technical sulfur, which was related to the higher bioavailability of the biogenic S^0 to the acidophiles (Ilyas and Lee, 2014b).

Mäkinen et al. (2015) studied the bioleaching of PCB froth, applying a pretreatment technique to separate the hydrophilic and hydrophobic fraction of crushed boards. The bioleaching medium consisted of 10 g/L S^0 and 4.5 g/L Fe^{2+}, along with various trace elements operated in a 3-L CSTR in three-step batch mode. A pre-cultivation method was followed to favor the dominance of sulfur-oxidizers over iron oxidizers in the bioleaching community. The pre-cultivation produced a bioleaching solution with pH 1.1, Fe^{3+} concentration of 7.4 g/L (Fe^{2+}

concentration was 0.4 g/L) and redox potential of +655 mV. When 50 g/L of PCB froth was added, the pH rose instantly, but was maintained at 1.6 with a H_2SO_4 (95%, v/v) addition of 250 mL/kg of PCB froth. Simultaneously, the redox potential dropped to +290 mV and the Fe^{2+} concentration increased to 6.8 g/L. During the next four days of bioleaching, there was a steady increase in redox potential and decrease in Fe^{2+} concentration, illustrating that some iron oxidation was occurring. With these parameters, Cu solubilization of 99% was reached in three days, with Cu and Fe being the only major metallic species in solution, corresponding into a maximal copper concentration of 6.8 g/L.

Chen et al. (2015) investigated column bioleaching of Cu from PCB that contained 24.8% Cu by *Acidithiobacillus ferrooxidans*. After column bioleaching for 28 days, the copper recovery reached at 94.8% from the starting materials. The study indicated that the copper dissolution rate is influenced by external diffusion rather than internal diffusion. As the Fe hydrolysis and formation of jarosite precipitates occur at the surface of the material, the bioleaching efficiency decreases. The formation of jarosite precipitates can be prevented by adding dilute H_2SO_4 and maintaining acidic conditions of the leaching medium. This enables the coupled cyclic Fe^{2+} - Fe^{3+} redox conversion and create optimal conditions for Cu bioleaching.

Conclusively, bioprocessing of WEEE for metal recovery from electronic waste is still at its infancy. Recent research work has focused on the improvement of two process parameters: (i) the leaching kinetics and (ii) the waste load in stirred tank bioreactors. Should these two parameters be improved, bioprocessing of waste material for metal recovery can become a competitive technology. The application of a combination of biological strategies could lead to higher solubilization efficiencies for many more elements compared to a singular approach. Lastly, a hybrid approach could be considered that combines the strength of both chemical and biological approaches.

2.9 Conclusions and perspectives

WEEE generation reached 42 Mt in the world in 2014, out of which 10.2, 9.3 and 6.0 Mt tons belonged to the EU-28, USA and China, respectively. It is the fastest growing type of domestic waste and its management requires special attention. Approximately 35% of the WEEE is reported to be collected in the EU, with a great variation between countries. Inappropriate management of WEEE is of environmental concern due to its hazardous nature. The rapid increase of electrical and electronic equipment (EEE) production and consequent WEEE generation are reliant on the availability of the required raw materials. Many of them are critical due to their limited supply, potential usage in other applications and economic importance. They are mostly found in complex mixtures and in their metallic/alloy forms. Thus, their recovery requires special attention in order to achieve selectivity.

WEEE is a very promising secondary source of critical metals. Specifically, printed circuit boards (PCB) are an important secondary source of Cu and Au, and platinum group metals (PGM), and the concentration of these metals are very high in certain components of WEEE. Additionally, the content of Ni, Zn, Al, are considerably high in discarded PCB. Hard disc drives (HDD) are an important source of REE, specifically Nd, Pr, and Dy owing to their high concentration in their permanent magnets. Displays are an important secondary source of indium (In), a critical raw material, along with aluminum (Al) and tin (Sb).

Conventional solid waste management practices are either inefficient and/or inappropriate for such a complex and potentially hazardous type of waste. Various physical, chemical and biological techniques have been proposed for valuing the waste as a secondary source of raw materials. There is no 'one-fits-all' strategy to recover valuable metals and to manage WEEE sustainably. In a circular economy, material loops are closed by recycling of discarded products, urban mining of EoL products and processing of current and future urban waste streams. State-of-the-art recycling technologies utilize a combination of physical technologies, where a fraction of WEEE is dismantled, crushed and subjected to thermal treatment. These technologies have limitations with regard to their environmental profile, and metal selectivity. Hydrometallurgy and increasingly biohydrometallurgy enables environmentally sound, metal selective and cost-effective processes to recover metals from waste material. Where necessary, novel methods could be combined with conventional methods, potentially creating novel hybrid technologies. Novel metal and unit specific recovery technologies, where a combination of various methods are expected to emerge in the coming years.

Chapter 3.

Biorecovery of metals from electronic waste – A review

This chapter is based on: Işıldar et al., 2016 Biorecovery of metals from electronic waste – A review, in book series Sustainable Technologies for Metal Recovery.

Abstract

Electronic waste, termed interchangeably as e-waste or waste electrical and electronic equipment (WEEE) is relatively the fastest growing segment of solid waste. The global electronic waste generation reached 42 million tons in 2014. In addition to being a highly hazardous waste type, electronic waste also includes relatively high concentrations of metals. Modern devices contain up to 60 metals at various concentrations. This encompasses base metals, critical metals, platinum group metals along with others, mixed in a complex matrix. Emergence of numerous new electronic products and occurrence of complex metals mixtures prove this waste stream to be an important secondary source of metals. Improper and informal end-of-life processing of electronic waste has detrimental consequences on the environment and public health. Microbial processing of metals from their ores is an established technology with many full-scale applications. Bioprocessing of waste materials to recover metals, on the other hand, is an emerging and promising biomass-based technology with low environmental impact and high cost-effectiveness. This chapter overviews bioprocessing of electronic waste as a secondary source of metals in order to recover metals. Additionally, biologically-driven metal extraction technologies, (e.g. bioleaching) and metal recovery techniques (e.g. biomineralization) are reviewed.

3.1 Introduction

The amount of discarded electric and electronic devices are growing at an increasing rate and the future trends show even larger amounts of electronic waste will be generated, in particular in growing economies (Wang et al., 2013). The global electronic waste generation reached 41.8 million tons in 2014, and forecasted to rise to 50 million tons in 2018 (Baldé et al., 2015). Although being highly toxic, electronic waste, in particular printed circuit boards (PCB) are promising secondary source of metals. It is expected that electronic waste, particularly via urban mining, will be an important secondary source of metals in the future.

PCB are Cu-dominated materials (approx. 20-25% by weight) along with a substantial amount of precious metals. Precious metals constitute the vast fraction of value of discarded PCB and are the main economic driver of metal recovery (Cui and Zhang, 2008). Metal recovery from electronic waste is not yet a fully established industry and currently there is considerable research work being carried out. There are several techniques for metal recovery, e.g.

pyrometallurgy (Yang et al., 2013), hydrometallurgy (Tuncuk et al., 2012) and biohydrometallurgy (Ilyas and Lee, 2015). Biomass-based processes, i.e. biohydrometallurgy encompasses a number of processes such as acidophilic (Liang et al., 2010) and cyanogenic bioleaching (Natarajan et al., 2015b), bioreduction (Yong et al., 2002), and biomineralization (Johnston et al., 2013). They have a number of advantages over analogous pyrometallurgical and hydrometallurgical processes such as being cost-effective and environmentally friendly in production of metals (Ilyas and Lee, 2014a).

Recent developments in biotechnology, such as acclimatization of microbes to extreme bioleaching conditions, indicate that biomass-based technologies could be a promising alternative to best available technologies. A comprehensive understanding of the metal mobilization mechanisms, toxicity characteristics, and process optimization would enable environmental biotechnology to play a major role in metal recovery from electronic waste. Recently, very high (99%) metal removal efficiencies (Mäkinen et al., 2015) at pulp densities up to 10% (Ilyas and Lee, 2014b) were achieved with bioleaching.

Biological approaches are often highlighted to play a significant role in the future of material processing for sustainable development. This applies not only to metal processing, but also the treatment of metal containing wastes and by-products (Lee and Pandey, 2012). An attractive feature of bioleaching is that it generates less pollutants compared to conventional metal processing. In this direction, an approach termed as 'process-integrated biotechnology' for a circular green economy has been propagated (Arundel and Sawaya, 2009). The importance of biotechnology is likely to increase in the future as high grade ore deposits are depleted. In this direction bioleaching is expected to become increasingly an integral part of metal processing, not only for primary sources but also for secondary sources.

3.2 Microbial mobilization of metals from electronic waste

3.2.1 Extraction of metals through biologically mediated reactions

Microbially mediated mobilization of metals, termed as bioleaching, is the conversion of metals from their solid form to water soluble forms, is an integral process in biohydrometallurgy. Extraction of metals from their ores takes place in the presence of microorganisms that are native to these environments (Brierley and Brierley, 2013). Biohydrometallurgy includes bioleaching and biorecovery processes, where aspects of environmental microbiology, biotechnology, hydrometallurgy, environmental engineering, mineralogy, and mining engineering merge.

Naturally occurring ores are processed predominantly via conventional methods. Pyrometallurgy, i.e. thermal treatment of ores, was replaced by modern hydrometallurgy at the end of the 19th century when two major operations were discovered: **(1)** the cyanidation process (MacArthur-Forrest process) for precious metals and **(2)** the Bayer process for refining bauxite, the primary Aluminum ore (Habashi, 2005). Biohydrometallurgy, on the other hand, is considered to have begun with the identification of the acidophile *Acidithiobacillus ferrooxidans* (reclassified from *Thiobacillus ferrooxidans*) as part of the microbial community found in acid mine drainage (Colmer and Hinkle, 1947). The first patent for a bioleaching process was granted in 1958 to the Kennecott Mining Company, showing the involvement of *At. ferrooxidans* for Cu extraction from low grade ore (Zimmerley et al., 1958). The patent describes a process where a leaching solution of ferric sulfate ($FeSO_4$) and sulfuric acid (H_2SO_4) is used. Ferric iron (Fe^{3+}) is regenerated by iron-oxidizing microorganisms through oxidation of ferrous iron (Fe^{2+}), and reused in a next leaching stage making the reaction cyclic. The biochemistry of bioleaching is explained further in detail in sections 3.2.3.3 and 3.2.4. Following the detection of *At. ferrooxidans* in the leachates in 1961, Rio Tinto mines in the Iberian peninsula have been among the first large-scale operations in which microorganisms played a major role (Brandl, 2008). Commercial application of biohydrometallurgy was effectively initiated in 1980 by the Lo Aguirre mine in Chile (Olson et al., 2003). The mine operated between 1980 and 1996 with a capacity of about 16,000 tons/day. This operation was followed by the emergence of a number of full-scale plants (Brierley and Brierley, 2001) and an increasing role of bioleaching plants is prevalent in the mining industry.

Today, bioleaching is increasingly used on a commercial scale for production of base metals, e.g. copper (Cu), nickel (Ni), zinc (Zn), molybdenum (Mo), cobalt (Co), lead (Pb), and metalloids, e.g. arsenic (As), gallium (Ga), antimony (Sb) in their sulfide and oxide ores, as well as the platinum group metals, e.g. platinum (Pt), rhodium (Rh), rubidium (Ru), palladium (Pd), osmium (Os), and iridium (Ir) associated with sulfide minerals (Brierley and Brierley, 2013). In case of Cu, an increasing number of full-scale bioleaching plants have started operated in the last decades (Schlesinger et al., 2011). On the other hand, recovery of metals from secondary sources using microbes is an emerging field of research. Some examples of metal-rich waste that could be regarded as secondary source of metals are mine waste (Liu et al., 2007), slags (Yin et al., 2014b; Potysz et al., 2016), sludges (Chen and Huang, 2014), contaminated soils (Deng et al., 2013), fly ashes (Ishigaki et al., 2005), spent catalyst (Lee and Pandey, 2012), and electronic waste (Hong and Valix, 2014).

3.2.2 Principles and mechanisms of microbial leaching

Various biological processes including bioleaching (microbially catalyzed leaching of metals), biooxidation (oxidation and enrichment of minerals by microorganisms), bioweathering (organic transformation of rocks and minerals over long time), and bioreduction (microbially induced reductive precipitation of metals in aqueous solutions) alter the chemistry and morphology of natural minerals. Acidophilic microorganisms thrive in low pH environments where microbial oxidation of minerals, e.g. pyrite (FeS), generate sulfuric acid (Rohwerder et al., 2003), resulting in the formation of acid mine drainage (Leff et al., 2015). Acidophiles are physiologically very diverse, spanning across aerobic and facultative anaerobic chemolithotrophs, various types of heterotrophic prokaryotes, as well as photoautotrophic eukaryotes (Xie et al., 2007). Mesophilic, thermophilic and hyperthermophilic species are commonly found in bioleaching environments (Rawlings and Johnson, 2007).

Acidophilic microorganisms keep their intracellular pH close to neutrality and maintain a proton gradient over their cytoplasmic membranes (Van de Vossenberg et al., 1998). Extracellular enzymes of acidophiles are optimally active at low pH (Bonnefoy and Holmes, 2012). Iron- and sulfur-oxidizer acidophiles are found in low pH environments and gain energy by oxidation of ferrous iron and inorganic sulfur compounds (Sand et al., 2001). They are also found in natural waters, sewer pipes causing corrosion problems, caves forming snottites (a layer of biomass which hang from the walls and ceilings of caves), hydrothermal vents, and geysers (Rawlings and Johnson, 2007).

Biomining has progressed from rather uncontrolled dump leaching to processing of refractory ores in designed bioheaps (Olson et al., 2003). Also, stirred tank bioprocessing has been developed and commercialized to full-scale (Acevedo, 2000). In bioleaching of sulfide minerals, the microorganisms play a catalytic role to oxidize ferrous (Fe2+) to ferric (Fe3+) iron and elemental sulfur (S^0) to sulfate (SO_4^{2-}) generating acid (Watling, 2006). Dissolution of certain metal sulfides, yields thiosulfate as an intermediate, which is further oxidized to sulfuric acid (Vera et al., 2013). Most of these microorganisms use atmospheric carbon dioxide as their carbon source and grow chemolithoautotrophically (Kimura et al., 2011).

In addition to acidophilic chemolithotrophic microorganisms, heterotrophic neutrophilic cyanide-generating microbes are also involved in bioleaching (Kaksonen et al., 2014). Many ubiquitous microorganisms are known to generate cyanide under certain conditions. Cyanide is the general name for the compounds consisting of a carbon atom triple-bonded to a nitrogen atom. They have high affinity to bond transition metals. Some soil bacteria, e.g. strains of *Chromobacterium violaceum* (Faramarzi et al., 2004), *Pseudomonas fluorescens* (Campbell et al., 2001), *Pseudomonas aeruginosa* (Fairbrother et al., 2009), as well as a number of fungal species, e.g. *Pleurotus ostreatus* (Brandl and Faramarzi, 2006) and algae, e.g. *Chlorella vulgaris* (Mata et al., 2009), have the ability to produce cyanide, and are used in processing of metals.

3.2.2.1 *Dissolution mechanism of metal sulfides: the thiosulfate and polysulfide mechanisms*

In this section, dissolution mechanisms of metals from their primary ores (metal sulfides) are reviewed. Despite similarities with metal sulfides, the main mechanisms of bioleaching of metals from waste material are still not well understood. The main peculiarity of secondary resources, such as waste material, is the speciation of metals (Tuncuk et al., 2012). The metals are found in their elemental form, often in various alloys. Due to the difference in chemical composition, their dissolution mechanism also shows differences.

Dissolution of metal sulfides follows two different pathways: the thiosulfate and the polysulfide pathway (Rohwerder et al., 2003). In general, dissolution is achieved by a combination of acidic leaching (proton attack) and oxidation processes. The reaction pathway is determined by the mineral species (Vera et al., 2013). The reactivity of metal sulfides with the protons is a significant criterion. Acid-insoluble sulfides, such as pyrite (FeS_2) are attacked through the thiosulfate mechanism, which depends on the oxidative attack of Fe^{3+} in solution. Acid soluble sulfides, such as chalcopyrite ($CuFeS_2$), are degraded through the polysulfide mechanism, as shown in Figure 3-1.

Fe^{3+} ions extract electrons from the mineral and are thereby reduced to the Fe^{2+} form. Consequently, the mineral releases metal cations and intermediate sulfur compounds. Iron-oxidizing bacteria catalyze the recycling of Fe^{3+}/Fe^{2+} cycle. In the case of the thiosulfate mechanism (acid-insoluble metal-sulfide mineral), an additional attack is performed by protons. The sulfur compounds are oxidized by sulfur-oxidizing bacteria and also abiotically. In Figure 3-1, H_2SO_4 (polysulfide mechanism) and elemental sulfur (thiosulfate mechanism) are highlighted in boxes. These main reaction products accumulate in the absence of sulfur-oxidizers.

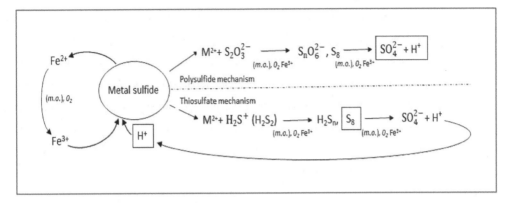

Figure 3-1: Thiosulfate and polysulfide leaching mechanisms of metal sulfide minerals by acidophiles, redrawn from Donati and Sand (2007).

Dissolution of metals from waste material was speculated to follow a similar mechanism (Ilyas et al., 2007; Liang et al., 2013). The metals are mostly found in their zero-valent state in waste material (Tuncuk et al., 2012), which makes the dissolution mechanism different. Elemental metals are mixed altogether, commonly in various complex alloys (Ongondo et al., 2015). Thus, addition of an external energy source is required. In addition, dissolution of metal sulfide minerals is an exothermic reaction, whereas it has not been observed for bioleaching of waste materials for metal recovery. Recent reports provided an insight into the bioleaching and metal solubilization mechanisms from waste materials, as discussed later in section 3.2.4.

3.2.2.2 *Physical contact mechanism: Contact, non-contact, and cooperative leaching*

Microbe-mineral interactions in bioleaching are explained by direct and indirect mechanisms (Watling, 2006). Metals are dissolved from minerals either directly by the metabolism of the cell or indirectly by the metabolic products. Direct mechanisms require close contact to the mineral, where microorganisms obtain electrons directly from the mineral, also termed as contact mechanism. In indirect or non-contact mechanism, the microorganisms are not attached to the surface. Instead they catalyze the oxidation of minerals by producing a leaching agent. In practice, a combination of both contact and non-contact mechanisms are involved, where the attached bacteria and the oxidizing agent in the solution play a role, termed cooperative leaching

(Rohwerder et al., 2003). The oxidation of the matrix is based on the activity of acidophilic chemolithotrophic iron-and sulfur-oxidizing microorganisms.

In the contact mechanism, a close contact is required where cells are attached to the mineral surface. It was shown that a significant fraction of the cells grows attached to the mineral using radioactively labeled carbon (^{14}C) *At. ferrooxidans* cells grown on $NaHCO_3$ (Escobar et al., 1996). The chemotactic behavior of *Leptospirillum ferrooxidans* to metal ions has been demonstrated (Acuna et al., 1992). Moreover, genes involved in chemotaxis were detected in *At. ferrooxidans* and *A. thiooxidans* (Valdés et al., 2008b). Contact bioleaching of the mineral could occur even in the absence of ferric ions. This explains the bioleaching of iron-free sulfides through such a mechanism (Rohwerder et al., 2003).

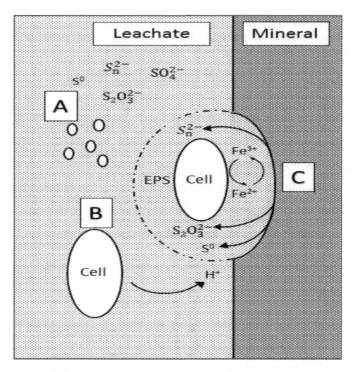

Figure 3-2: Schematic mechanisms of cooperative bioleaching; sulfur oxidation (a), non-contact (B), and contact (c) mechanisms, redrawn from Tributsch, (2001) and Brandl, (2008).

When minerals are bioleached, microorganisms metabolize lixiviants, either through direct electron transfer or indirectly, and create the space in which bioleaching take place. The extracellular polymeric substances (EPS) may serve as the reaction space and many complex bioleaching reactions take place within this EPS layer, rather than in the solution (Sand et al., 2001). The space between the cell wall and the surface is considered as reaction space. Many species typically form biofilms from an EPS layer when they attach to the surface of a mineral (Ghauri et al., 2007). Bioleaching of metals from WEEE shows a similar pattern, as explained later in 3.2.4.

3.2.3 Metal mobilization mechanisms

A number of metal mobilization mechanisms are defined, namely (i) acidolysis (formation of acids), (ii) complexolysis (excretion of complexing agents), and (iii) redoxolysis (microbially induced or catalyzed oxidation and reduction reactions) (Bosecker, 1997; Brandl, 2008).

3.2.3.1 *Acidolysis*

In the acidolysis mechanism, the dissolution of metals occurs by biogenic acids (Vera et al., 2013). Microorganisms catalyze the protonic mechanism, in which they excrete protons that weaken the metal ion bond, thus bringing the metal into solution. In most cases, mineral solubilization occurs simultaneously in the presence of the metabolic ligands under acidic conditions (Brandl and Faramarzi, 2006). In bioleaching of metals from waste materials, the prerequisite is that the bonds between metal ions and ligands are stronger than those between metal ions and solid particles. In that case, the metal is successfully leached from the solid particles into the solution. Acidolysis is performed by a number of autotrophic sulfur oxidizers and heterotrophic fungal and bacterial cells. A list of microorganisms that can perform acidolysis are given in Table 3-1.

3.2.3.2 *Complexolysis*

In the complexolysis mechanism, metal solubilization is induced by a metabolic ligand which increases the metal mobility by complexation and/or chelation. Biogenic complexing agents bind with metals, replacing bonds, leading to detachment of metals into solution. Siderophores, secreted by a number of bacteria and fungi are amongst the strongest soluble iron, (Fe^{3+}) binding agents known (Rawlings, 2005). Fe^{3+} is found mostly insoluble in natural waters. Along with their water soluble Fe^{3+} iron chelating ability, siderophores can also bind other metals (Neilands, 1995; Del Olmo et al., 2003).

The complexolysis mechanism is largely performed by heterotrophic cyanide-generating microorganisms. Cyanide is the general term for chemicals which contain a cyano group with the chemical formula CN^- that occur naturally or artificially. Humans have trace amounts of thiocyanide (SCN) in saliva, urine, and gastric juices (Zammit et al., 2012). CN^- reacts with metals in the waste material as a complexing agent and forms soluble metal-cyanide complexes

(Rees and Van Deventer, 1999). In particular, recovery of noble metals from secondary sources is focused on the utilization of cyanogenic bacteria. Precious metals such as Au, Pt, and Pd are among the most chemically stable elements, and react with only a limited number of chemicals. Safety issues regarding cyanide can be minimized because cyanogenic bacteria autonomously decompose cyanide to nontoxic β-cyanoalanine (Knowles, 1976). Thus, the biological cyanide production process enables the design of a system without need of an additional treatment (Shin et al., 2013). Cyanide-complexed metals can be subsequently recovered using various methods, such as adsorption, electrowinning, or cementation.

3.2.3.3 *Redoxolysis*

In the redoxolysis mechanism, microorganisms produce catalytic compounds which regulate the oxidation potential of the solution. Leaching efficiency and rate depend on the mineral phase, type of metal and oxidation state (Mishra and Rhee, 2014). Fe^{3+} is one of the most common redoxolysis agents in leaching systems. It is produced by iron oxidizers and is reduced to Fe^{2+} in the bioleaching reaction, re-oxidized by iron oxidizers to Fe^{3+}, making the reaction cyclic (Schippers et al., 1996).

The thermodynamic equilibrium of reactions with Cu, Zn, and Ni can be elucidated on the basis of enthalpy and Gibbs free energy values under normal conditions, given below in equations (3-1), (3-2), and (3-3).

$$Cu^0 + 2Fe^{3+} \rightarrow Cu^{2+} + 2Fe^{2+} (G^0 = -347.1 \text{ kJ}) \tag{3-1}$$

$$Ni^0 + 2Fe^{3+} \rightarrow Ni^{2+} + 2Fe^{2+} (G^0 = -822.6 \text{ kJ}) \tag{3-2}$$

$$Zn^0 + 2Fe^{3+} \rightarrow Zn^{2+} + 2Fe^{2+} (G^0 = -1235.9 \text{ kJ}) \tag{3-3}$$

Fe^{3+} act as oxidizing agent, and readily oxidizes metals leading to their dissolution. It has a standard reduction potential of 0.77 V. Redox potential is an essential parameter highly useful to estimate and understand the chemistry and speciation of iron (Yue et al., 2016). In bioleaching environments, an increased redox potential is observed due to the activity of iron-oxidizing microorganisms. Also, many microbial strains have the ability to reduce Fe^{3+} to Fe^{2+} under anaerobic conditions (Rawlings, 2005). The redox chain from Fe^{2+} ions to the final electron acceptor oxygen has been shown for mesophilic acidophilic iron oxidizers (Brasseur et al., 2004).

3.2.4 Bioleaching of metals from electronic waste material

Reported Cu bioleaching efficiencies vary widely from 50% to 100% with leaching periods typically exceeding 5 days and pulp densities of 1-3% (*w/v*). Several studies have demonstrated improved bioleaching efficiencies in sulfur- and ferrous iron-supplemented media (Wang et al., 2009; Ilyas et al., 2010; Liang et al., 2010). A summary of the recent literature is given in Table 3-2.

From a recycling point of view, the main peculiarity of the electronic waste material is that the metals are found in their zero-valent elemental state, present in complex alloys (Tuncuk et al., 2012). This requires novel strategies to effectively recover metals from waste material. Metal mobilization from waste materials by acidophiles involves an indirect leaching mechanism by biogenic reagents such as H_2SO_4 and Fe^{3+} produced in the first stage. In addition, the importance of bacterial attachment to electronic waste material has been recently demonstrated: *At. ferrooxidans* showed a 25% lower Cu mobilization efficiency from PCB when the contact of bacterial cells and the crushed electronic was intentionally avoided by a selective membrane (Silva et al., 2015).

The mechanism of metal dissolution from electronic waste material has been long debated. It has been speculated that the mechanism of Cu leaching from printed circuit boards by *At. ferrooxidans* was similar to that of metal sulfides (Ilyas et al., 2010). It may involve indirect leaching mechanisms by the biogenic sulfuric acid, where the role of the microorganisms is to oxidize elemental sulfur (S^0) to sulfuric acid (H_2SO_4). Fe^{2+} in the solution plays a role as an electron donor, and is oxidized to Fe^{3+} by iron-oxising bacteria. Fe^{3+} then plays a acts as an oxidizing agent as shown in equation (3-5) catalyzing the leaching reaction. This translates into a combined acidolysis – redoxolysis bioleaching mechanism for metal dissolution from waste materials, as shown below in equations (3-4) and (3-5):

$$4Fe^{2+} + O_2 + 2H^+ \rightarrow 4Fe^{3+} + 2OH^- \qquad \textit{(bacteria)} \qquad (3\text{-}4)$$

$$S^0 + 1.5O_2 + H_2O \rightarrow 2H^+ + SO_4^{2-} \qquad \textit{(bacteria)} \qquad (3\text{-}5)$$

Biogenic Fe^{3+} and H_2SO_4 mobilizes Cu from electronic waste material as shown below in, respectively, equations (3-6) and (3-7).

$$Cu^0 + 2Fe^{3+} \rightarrow Cu^{2+} + 2Fe^{2+} \qquad \textit{(chemical)} \qquad (3\text{-}6)$$

$$Cu^0 + 2H^+ + SO_4^{2-} + 0.5O_2 \rightarrow Cu^{2+} + SO_4^{2-} + H_2O \qquad \textit{(chemical)} \qquad (3\text{-}7)$$

Metal removal efficiency reportedly decreases significantly with increasing pulp density. This has several possible causes. Firstly, the waste material has an alkaline nature, and therefore is acid-consuming (Brandl et al., 2001). This results in a high pH environment, where the acidophiles do not thrive. Secondly, the non-metallic fraction, i.e. epoxy-coated substrate, organic fraction etc., of the material is toxic to the bacteria (Niu and Li, 2007; Zhu et al., 2011).

The Cu mobilization rate primarily depends on the initial pH, Fe^{2+} concentration and oxidation rate of Fe^{2+} to Fe^{3+} ions generated by iron oxidizers (Xiang et al., 2010). The biogenic Fe^{3+} concentration is correlated with the metal mobilization rate and leaching efficiency (Zhu et al., 2011). This confirms the involvement of the indirect leaching mechanism by sulfuric acid and ferric iron in mobilizing metals. On the other hand, the involvement of contact mechanism, the physical attachment of the bacterial cells to the electronic waste material, is strongly mediated by the ionic strength of the solution. *At. ferrooxidans* cells do not attach randomly to the solid surface, but chemotaxis could be involved in the preferential attachment of bacteria (Rohwerder et al., 2003). The interaction between *At. ferrooxidans* cells and crushed PCB particles is favorable only if the van der Waals attractive force is greater than the electrostatic repulsive force. This greatly depends on the ionic strength of the solution.

In a recent study, the involvement of both contact and non-contact mechanisms was shown, where the final fraction of Cu mobilized was significantly lower (25%) in a system where contact mechanism was avoided: Ground PCB sample of particle size 500 - 1000 µm was placed inside a semi-permeable Molecular Weight Cut Off (MWCO) membrane partition system, to study the effect of the contact leaching mechanism on the final leaching efficiency. The experimental setup is shown in Figure 3-3. The main finding was that the bacterial adhesion is responsible for the higher Cu extraction rate. The results for bacterial adhesion tests were consistent with the Derjaguin–Landau–Verwey–Overbeek theory (Silva et al., 2015) which explains the aggregation of aqueous dispersions and describes the force between charged surfaces interacting through a liquid medium. It combines the effects of the Van der Waals attraction and the electrostatic repulsion due to the so-called double layer of counter-ions.

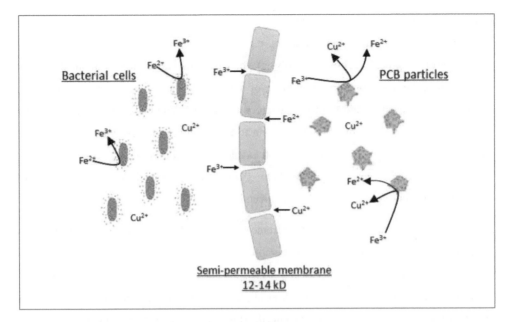

Figure 3-3: Bacterial adhesion model by *Acidithiobacillus ferrooxidans* on crushed PCB in bioleaching (Redrawn from Silva et al., 2015).

It was shown that 24% of the total cells (4.3×10^7 per gram) were attached on PCB in the initial 60 min. Bacterial attachment evidently plays a crucial role in Cu extraction efficiency from PCB. Furthermore, a decrease in the oxidation-reduction potential (ORP) was observed when bacterial contact was avoided. The higher ORP value of contact conditions over non-contact conditions suggests that the oxidation of Fe^{2+} by the attached bacteria occurs (Silva et al., 2015).

3.2.5 Microorganisms

Bioleaching of metals is performed by a diverse group of microorganisms. Native bacteria in natural ores and metal-rich environments grow on the surface of the metal bearing minerals (Ghauri et al., 2007). A variety of lithotrophic and organotrophic microorganisms mediate the leaching processes (Bharadwaj and Ting, 2011). Mainly, three groups of microorganisms are classified in bioleaching, namely (1) chemolithotrophic prokaryotes, including bacteria and archaea, (2) heterotrophic bacteria and (3) heterotrophic fungi are classified (Johnson and Du Plessis, 2015). The majority of the identified acidophiles belongs to the mesophilic and moderately thermophilic bacteria. Many acidophilic bioleaching archaea are thermophilic. Although thermophiles are reported to improve the reaction rates and extent of metal extraction due to elevated temperatures (Olson et al., 2003), most commercial applications are reported to be operated at mesophilic temperatures, i.e. below 40°C (Ilyas and Lee, 2015).

Sulfur oxidizers, e.g. *Acidithiobacillus thiooxidans,* generate sulfuric acid (acidolysis) which result in leaching of metals. Iron oxidizers, e.g. *Leptospirillum ferrooxidans* are involved in production of biogenic ferric iron (redoxolysis) which is a powerful oxidizer bringing the metals to their ionic form in solution. A number of soil bacteria and fungi excrete complexing compounds, e.g. cyanide and form chelates (complexolysis). The major microorganisms that are found in bioleaching environments belong to the genera *Acidimicrobium, Acidisphaera, Acidithiobacillus, Acidobacterium, Acidocella, Acidiphilium, Alicyclobacillus, Ferrimicrobium, Frateuria, Leptospirillum, Sulfobacillus,* and *Thiomonas.* A list of microorganisms involved in bioleaching is given in Table 3-1.

Acidophiles are able to readily adapt to extreme conditions (Hedrich et al., 2011) which is also observable in their vast genetic variation among many strains (Kimura et al., 2011). The full genome of several strains of bioleaching organisms such as *At. ferrooxidans* (Valdés et al., 2008a), *Metallosphaera sedula* (Auernik et al., 2008) and cyanide-producing *Pseudomonas putida* (Canovas et al., 2003) were sequenced, which gave an insight into the genes enabling bioleaching mechanisms, heavy metal resistance and cell-metal interactions. Moreover, gene modification of well-studied organism *Chromobacterium violaceum* has been experimented so as to enhance the microbes ability to produce metal complexing metabolites (Natarajan et al., 2015b).

Table 3-1: Microorganisms involved in bioleaching.

Microorganism	Leaching mechanism	Temperature optimum	pH optimum	References
Archaea				
Acidianus brierleyi	Acidolysis (H_2SO_4), redoxolysis (Fe^{3+})	Thermophilic (70°C)	1.5 - 2	Brierley and Brierley, 1986; Nemati and Harrison, 2000
Ferroplasma acidarmanus	Acidolysis (H_2SO_4), redoxolysis (Fe^{3+})	Moderately thermophilic (42°C)	0.5 - 1.2	Edwards et al., 2000
Ferroplasma cupricumulans	Redoxolysis (Fe^{3+})	Moderate thermophilic (54-63°C)	1.0 - 1.2	Hawkes et al., 2006
Ferroplasma acidiphilum	Redoxolysis (Fe^{3+})	Mesophilic (15-45°C)	1.3 - 2.2	Golyshina et al., 2000
Metallosphaera hakonensis	Acidolysis (H_2SO_4)	Thermophilic (55-80°C)	1.0 – 4.0	Plumb et al., 2008
Metallosphaera sedula	Acidolysis (H_2SO_4)	Thermophilic (75°C)	1.0 - 4.5	Huber et al., 1989

Microorganism	Leaching mechanism	Temperature optimum	pH optimum	References
Metallosphaera sedula	Acidolysis (H_2SO_4), redoxolysis (Fe^{3+})	Thermophilic (75°C)	1.0 - 4.5	Auernik et al., 2008
Metallosphaera yellowstonensis	Acidolysis (H_2SO_4), redoxolysis (Fe^{3+})	Thermophilic (65°C)	2.5 - 3.5	Kozubal et al., 2011
Sulfolobus acidocaldarius	Acidolysis (H_2SO_4), redoxolysis (Fe^{3+})	Extreme thermophilic (55-85°C)	2.0 – 3.0	Plumb et al., 2002
Sulfolobus metallicus	Acidolysis (H_2SO_4), redoxolysis (Fe^{3+})	Thermophilic (50-75°C)	2.0 – 3.0	Brandl et al., 2008; Kaksonen et al., 2014
Sulfolobus yangmingensis	Acidolysis (H_2SO_4)	Extreme thermophilic (80°C)	4.0	Jan et al., 1999
Thermoplasma acidophilum	Acidolysis (H_2SO_4), redoxolysis (Fe^{3+})	Moderately thermophilic (55-60°C)	1.0 – 2.0	Darland et al., 1970; Ilyas et al., 2010

Bacteria

Microorganism	Leaching mechanism	Temperature optimum	pH optimum	References
Acidimicrobium ferrooxidans	Redoxolysis (Fe^{3+})	Moderately thermophilic (45-50°C)	2.0	Clark and Norris, 1996

Microorganism	Leaching mechanism	Temperature optimum	pH optimum	References
Acidithiobacillus caldus	Acidolysis (H_2SO_4)	Moderately thermophilic (42-45°C)	2.0 - 2.5	Zhou et al., 2007
Acidithiobacillus ferrooxidans	Acidolysis (H_2SO_4), redoxolysis (Fe^{3+})	Mesophilic (28-35°C)	2.5	Silverman and Lundgren, 1959; Kelly and Wood, 2000
Acidithiobacillus thiooxidans	Acidolysis (H_2SO_4)	Mesophilic (28-30°C)	1.0 – 3.0	Waksman and Joffe, 1922; Bosecker, 1997
Ferrimicrobium acidiphilum	Redoxolysis (Fe^{3+})	Mesophilic (35°C)	1.4 - 2.0	Johnson et al., 2009
Ferrithrix thermotolerans	Redoxolysis (Fe^{3+})	Moderately thermophilic (43-50°C)	1.6 - 1.8	Johnson et al., 2009
Ferrovum myxofaciens	Redoxolysis (Fe^{3+})	Mesophilic (20-30°C)	1.0 – 2.0	Fabisch et al., 2013
Leptospirillum ferriphilum	Redoxolysis (Fe^{3+})	Moderately thermophilic (42°C)	1.2 - 1.6	Spolaore et al., 2011

Microorganism	Leaching mechanism	Temperature optimum	pH optimum	References
Leptospirillum ferrooxidans	Redoxolysis (Fe^{3+})	Mesophilic (28-35°C)	1.8	Sand et al., 1992
Leptothrix discophora	Redoxolysis (Fe^{3+})	Mesophilic (15-40°C)	5.8 - 7.8	Corstjens et al., 1992
Chromobacterium violaceum	Complexolysis (CN^-)	Mesophilic (25-37°C)	7.0 - 7.5	Campbell et al., 2001
Pseudomonas aeruginosa	Complexolysis (CN^-)	Mesophilic (25-35°C)	7.0 - 7.8	Castric, 1977
Pseudomonas fluorescens	Complexolysis (CN^-)	Mesophilic (25-35°C)	7.0 - 7.8	Blumer and Haas, 2000
Pseudomonas putida	Acidolysis (citrate), complexolysis (CN^-)	Mesophilic (25-35°C)	7.0 - 8.2	Brandl and Faramarzi, 2006
Sulfobacillus sibiricus	Acidolysis (H_2SO_4), redoxolysis (Fe^{3+})	Moderately thermophilic (45-55°C)	1.7 – 2.0	Melamud et al., 2003
Sulfobacillus thermosulfioxidans	Acidolysis (H_2SO_4), redoxolysis (Fe^{3+})	Moderately thermophilic (40-60°C) (50°C)	1.7 – 2.0	Golovacheva and Karavaiko, 1978

Microorganism	Leaching mechanism	Temperature optimum	pH optimum	References
Fungi				
Aspergillus awamori	Acidolysis (citrate), complexolysis (Oxalate)	Mesophilic (28°C)	6.8 - 7.2	Mapelli et al., 2012
Aspergillus flavus	Acidolysis (organic acids)	Mesophilic (30°C)	6.5	Mishra et al., 2009
Aspergillus fumigatus	Acidolysis (citrate), complexolysis (Oxalate)	Mesophilic (30°C)	6.6 - 7.2	Brandl, 2008
Aspergillus niger	Acidolysis (citrate, oxalate), complexolysis (CN⁻)	Mesophilic (30°C)	6.4 - 7.3	Xu and Ting, 2009
Cladosporium oxysporum	Acidolysis (organic acids)	Mesophilic (30°C)	6.5	Mishra et al., 2009

3.2.5.1 *Chemolithoautotrophs*

Certain groups of prokaryotes obtain their energy from the oxidation of reduced inorganic compounds, and are called chemolithotrophs. The majority of acidophilic bioleaching organisms are autotrophs that use inorganic carbon (CO_2) as carbon source (Donati and Sand, 2007). They derive energy from the oxidation of inorganic compounds such as Fe^{2+} or reduced sulfur compounds, such as elemental sulfur (S^0) and metal sulfides (MeS). Some species also derive energy from the oxidation of hydrogen gas under aerobic and anaerobic conditions (Hedrich and Johnson, 2013). Bacterial leaching is carried out in an acidic environment at low pH ranging between 0.5 and 4.0 when most metal ions remain in solution. Most chemolithoautotrophs have high tolerance for heavy metals (Orell et al., 2010). *Acidithiobacillus ferrooxidans, Acidithiobacillus thiooxidans* and *Leptospirillum ferrooxidans* are the most extensively studied mesophilic microbes in bioleaching communities. There is an increasing interest in thermophilic chemolithoautotrophic bioleaching with involvement of microorganisms such as *Acidianus brierleyi, Sulfobacillus thermosulfidooxidans* and *Metallosphaera sedula* (Du Plessis et al., 2007). These acidophiles grow on iron-and sulfur-containing mining ores such as pyrite (FeS_2), pentlandite ($(Fe,Ni)_9S_8$) and chalcopyrite ($CuFeS_2$) at temperatures in the range of 45-75°C.

3.2.5.1.1 Genus *Acidithiobacillus*

The genus *Acidithiobacillus* belong to γ-proteobacteria and is considered as one of the most important group of microorganisms in biomining. These bacteria are obligate acidophilic, gram-negative rods sized in average 0.4×2.0 µm (Figure 3-4), motile by one or more flagella, using iron and sulfur for autotrophic growth. They exhibit very high genetic variation (Kimura et al., 2011). Formerly known as *Thiobacillus,* many species including *Thiobacillus thiooxidans,* and *Thiobacillus caldus,* as well as *Thiobacillus ferrooxidans* were reassigned to genus *Acidithiobacillus* on the basis of physiological characters and 16S rRNA gene sequence comparisons (Kelly and Wood, 2000). *Acidithiobacillus* and especially the bacterium *Acidithiobacillus ferrooxidans* is a major microbe in bioleaching communities (Johnson and Hallberg, 2003). It uses atmospheric sources for both its carbon and nitrogen. This species plays an important role in the biogeochemical cycling of metals in the environment, being involved in solubilizing of minerals and also immobilizing of metal cations (Gadd, 2010). As with many other chemolithotrophs, their ability to oxidize Fe^{2+}, and less commonly S^0, is the key characteristic of the species of the genus *Acidithiobacillus*. These bacteria are abundant in natural environments associated with pyritic ore bodies, coal deposits, and acid mine drainages. Many strains are potentially agents to extract and recover metals, as well as to assist in bioremediation applications. Research work on autotrophic mesophilic bioleaching of metals from waste materials, more specifically discarded PCB, has been focused on the species of the genus *Acidithiobacillus*, especially *A. ferrooxidans*, and *A. thiooxidans* (Wang et al., 2009; Hong and Valix, 2014).

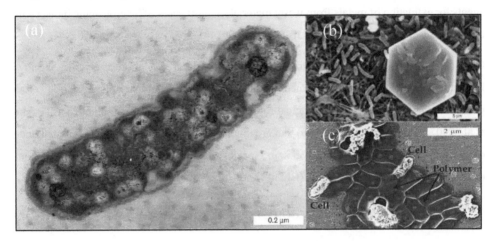

Figure 3-4: Transmission electron microscopy (TEM) micrographs of single cell *Acidithiobacillus ferrooxidans* (A), bioleaching of copper and iron from molybdenite (MoS$_2$) ore (B) and attached to pyrite surface surrounded by polymer (C) (Sources: Murr et al., 2014; Romato et al., 2001; Edwards et al., 2001).

At. ferrooxidans is one of the few microorganisms known to gain energy by the oxidation of Fe^{2+} in acidic environments, generating reverse electron flow from Fe^{2+} to NADH (Yin et al., 2014a). Its importance in industrial applications led to complete sequencing of the genome (Valdés et al., 2008a). The iron and sulfur oxidation mechanisms, nutrient uptake, heavy metal resistance mechanisms, biofilm formation, quorum sensing, inorganic ion uptake of this microorganism are explained in detail by Valdés et al. (2008a). *At. thiooxidans* is a mesophilic obligate aerobe that couples the oxidation of elemental sulfur and a variety of reduced sulfur compounds to sulfate, coupled with production of protons. It is one of the first acidophilic microorganism that was isolated in 1922 by Waksman and Joffe (Bosecker, 1997) capable of growing at pH 0.5.

3.2.5.2 *Heterotrophs*

Members of the acid- and complexant-producing bacterial genera *Bacillus, Pseudomonas, Chromobacterium* as well as fungal genera *Aspergillus, Penicillium* are extensively studied in bioleaching. Compared to acidophiles, heterotrophs tolerate a wider pH range and are employed for treating moderately alkaline wastes (Natarajan et al., 2015a). Research work on heterotrophic bioleaching of metals from waste materials has been focused on the cyanide-and organic acid-generating microorganisms. Heterotrophic bacteria and fungi are involved in bioleaching with microbial production of organic acids (Bosecker, 1997). Organic acids play a role as bioleaching agents (Brandl, 2008). Also, other metabolites could play a role as leaching agents for extraction of metals from waste material. In most cases of heterotrophic bioleaching, organic acids directly solubilize metals (Gadd, 2000).

Cyanide is a secondary metabolite formed by oxidative decarboxylation of glycine, as shown in equation (3-8). It is typically formed only during the early stationary phase, and in certain growth media. Cyanide has an ecological role e.g. suppressing diseases on plant roots (Bakker et al., 2007). Induction of genes (*hcn*) that are involved in cyanide production is initiated under oxygen limitation conditions, however some species are known to produce reasonable amounts of cyanide under normal conditions (Blumer and Haas, 2000).

$$NH_2CH_2COOH \rightarrow HCN + CO_2 + 2H_2 \tag{3-8}$$

Biological production of cyanide by bacteria depends on a number of fundamental parameters such as precursor concentration (i.e. glycine), initial pH, operating temperature, presence of waste material/ore, and oxygen concentration (Fairbrother et al., 2009; Shin et al., 2013). Of the above-listed parameters, especially precursor concentration and pH have a narrow optimum range. Glycine is essential for biogenic cyanide production however, high concentrations of glycine is reported to be inhibitory for growth (Castric, 1977). CN^- is chemically stable at high pH, and gold cyanidation is most efficient in the range of 10.5-11. However a pH higher than 9.5 is inhibitory for cell growth for most cyanogenic bacteria (Liang et al., 2014). A few studies investigated the adaptation of cyanogenic bacteria to pH values above 9.5 so as to promote the leaching efficiency (Ting and Pham, 2009; Natarajan and Ting, 2014). *Chromobacterium violaceum*, which is the most widely studied cyanogenic heterotroph despite several strains being opportunistic pathogens, can adapt to pH values up to 9.5. Adapted cells enabled increased bioleaching efficiency due to increased chemical stability of CN^- under alkaline conditions.

3.2.5.2.1 Genus *Pseudomonas*

The Genus *Pseudomonas* encompasses some of the most well-studied and versatile heterotrophs in biotechnology in a wide field of applications. A number of strains such as *P. aeruginosa*, *P fluorescens*, *P. putida*, are commonly utilized in bioleaching of metals. They are ubiquitous microbes, typically found in soil biota, and significant in bioleaching due to their various metabolites. Cyanide is optimally excreted during growth limitation and may provide the microbe, which is usually cyanide-tolerant, a selective advantage (Kaksonen et al., 2014). Cyanide mechanisms may occur in rhizosphere soils with tops rich in organic matter, where a potential symbiotic relationship between the plants cyanide excreting microorganisms is speculated (Ubalua, 2010).

The exact mechanism of the biological Au solubilization is not yet explained. It is shown to be a complexolysis reaction between Au and catalytic cyanide (Fairbrother et al., 2009). Au recovery by *Pseudomonas* is experimented by many researchers on primary ores (Shin et al., 2013) and crushed WEEE (Pradhan and Kumar, 2012). Several strategies including sequential nutrient addition (Brandl et al., 2008), genetic modification (Natarajan et al., 2015b), and medium modification (Natarajan and Ting, 2015b) are developed to increase the cyanide production of the microorganisms.

3.2.5.2.2 Fungi

Fungal bioleaching occurs through acidolysis and redoxolysis mechanisms, via metabolic citric acid, oxalic acid, and gluconic acid (Deng et al., 2013). These acids induce the leaching of metals from ores and waste materials by regulating redox potential and acidity (Ubaldini et al., 1998; Gadd, 2010). In contrast to acidophilic bacterial leaching, fungal bioleaching takes place at a relatively higher pH (Xu and Ting, 2004). *Aspergillus niger* and *Penicillium simplicissimum* are among the most used microbes in bioleaching of metals from waste material (Lee and Pandey, 2012). In presence of WEEE, they are able to adapt to high pulp densities up to 10% (*w/v*) in about 5 - 6 weeks (Brandl et al., 2001).

3.2.6 Bioreactors

Bioprocessing of metals from natural primary ores has developed into a successful technique with a number of full-scale reactors currently being operated (Brierley and Brierley, 2013). Currently, an increasing amount of the global Cu production, around 20%, is carried out by bioleaching plants (Schlesinger et al., 2011). Moreover, biooxidation of refractory concentrates, mostly of Au ores, are well-established full-scale processes which take place in tank reactors settings (Acevedo, 2000). However, for the secondary materials, such as WEEE, it is still at its infancy. Recent studies focused on increasing the waste load rate in reactors, and remediation of the toxic effect of electronic waste on the bacteria.

Ilyas et al. (2007) investigated column bioleaching of metals from electronic waste by moderately thermophilic acidophilic bacterial strains of chemolithotrophic and heterotrophic consortia. They reached a final leaching efficiency of 64%, 86%, 74%, and 80%, for, Al, Cu, Ni, and Zn, respectively, after a pre-leaching period of 27 days followed by a bioleaching period of 280 days. Follow up studies from the research group on a continuous stirred reactor (CSTR) bioleaching setup with moderately thermophilic adapted cultures of *Sulfobacillus thermosulfidooxidans* resulted in 91%, 95%, 94%, and 96% efficiency for Al, Cu, Ni, and Zn, respectively, at a pulp density of 10% (*w/v*). The stirred reactor was supplemented with enriched air of 25% O_2 + 0.03% CO_2, with 2.5% (*w/v*) biogenic S^0 and maintained at 45°C. An interesting finding was the faster oxidation rate of biogenic sulfur over technical sulfur, which can be attributed to the higher bioavailability and hydrophobicity of biogenic sulfur (Ilyas and Lee, 2014b).

Mäkinen et al. (2015) studied the bioleaching of PCB froth in a CSTR, applying a pretreatment technique to separate the hydrophilic and hydrophobic fraction of crushed boards. The bioleaching medium consisted of 10 g/L S^0 and 4.5 g/L Fe^{2+}, along with various trace elements operated in a 3-L CSTR in three-step batch mode. A pre-cultivation method was followed so as to favor the dominance of sulfur-oxidizers over iron oxidizers in the bioleaching community. The pre-cultivation produced a bioleaching solution with pH 1.1, Fe^{3+} concentration of 7.4 g/L (Fe^{2+} concentration was 0.4 g/L) and redox potential of +655 mV. When 50 g/L of PCB froth was added, the pH rose instantly, but was maintained at 1.6 with total sulfuric acid (95% *v/v* H_2SO_4) addition of 250 mL/kg of PCB froth. Simultaneously, the redox potential dropped to +290 mV and the Fe^{2+} concentration increased to 6.8 g/L. During the next four days of bioleaching, there was a steady increase in redox potential and decrease in Fe^{2+} concentration, illustrating that some iron oxidation was occurring. With these parameters, Cu solubilization of 99% was reached in three days, with Cu and Fe being the only major metallic species in solution, yielding a maximal Cu concentration of 6.8 g/L (Mäkinen et al., 2015).

Chen et al. (2015) investigated column bioleaching of Cu from PCB by *Acidithiobacillus ferrooxidans*. After column bioleaching for 28 d, the Cu recovery reached at 94.8% from the starting materials contained 24.8% Cu. The study indicated that the Cu dissolution rate is influenced by external diffusion rather than internal rate. As the iron hydrolysis and formation of jarosite precipitates occur at the surface of the material, bioleaching efficiency decreases. The formation of jarosite precipitates can be prevented by adding dilute sulfuric acid and maintaining an acidic condition of the leaching medium. This enables the coupled cyclic Fe2+– Fe^{3+} and create optimal conditions for Cu bioleaching.

Nie et al. (2015a) bioleached 96% of the Cu from discarded PCB extracted from metal concentrates by an *A. ferrooxidans*-dominated mixed culture in 7 days in a combined CSTR reactor. At an initial Fe^{2+} concentration of 12 g/L.with an average 0.2307 g L/h Fe^{2+} ion oxidation rate was measured culture of acidophilic bacteria. Protons produced by the ionization of sulfuric acid (acidolysis) and the hydrolysis of Fe^{3+} played only a slight role in the extraction of Cu. Dialysis bag experiments show 81.4% of Cu was leached out by bioleaching without dialysis bag compared with 47.9% in the encapsulated bioleaching system. The extraction of Cu was mainly through the indirect oxidation process (redoxolysis) via Fe^{3+} ions biogenerated by *At. ferrooxidans*. Both contact and non-contact mechanisms led to the dissolution of Cu from discarded PCB.

Despite its many advantages, bioleaching of waste material for metal recovery has a number of constraints. Bioleaching processes are limited by several factors such as lengthy leaching periods up to 15 days as well as toxic effect of the waste material on the microorganisms. In this direction, there is vast potential for process optimization, particularly with the optimization of the biological reactions. Recent studies showed that the metal leaching efficiency of 99% could be achieved with improved kinetics (three days) at relatively high pulp densities of 10% (*w/v*). Bacteria are able to tolerate conditions previously considered to be highly toxic after an adaptation period in reactor setups. Further engineering of this property allows development of novel biotechnology processes in the context of bioprocessing of electronic waste for metals recovery.

Table 3-2: Bioleaching of metals from PCB via various mechanisms.

Microorganism(s)	Leaching mechanism	Temp.°C	pH	Reactor type	Pulp density (w/v)	Leached metals (percentage, mg/g PCB)	References
Acidithiobacillus thiooxidans	Acidolysis (H_2SO_4)	30°C	0.5	Batch reactor	1%	**Cu** 98% (832 mg/g)	Hong and Valix, 2014
Aspergillus niger, Penicillium simplicissimum	Acidolysis (organic acids)	30°C	3.5	Batch reactor	1%	**Cu** 65% (52 mg/g), **Al** 95% (225 mg/g), **Ni** 95% (14 mg/g), **Zn** 95% (25 mg/g)	Brandl et al., 2001
Chromobacterium violaceum (metabolically engineered)	Complexolysis (CN^-)	30°C	Neutral	Batch reactor	0.5%	**Au** 31% (0.04 mg/g)	Tay et al., 2013
Sulfobacillus thermosulfidooxidans, acidophilic isolate	Acidolysis (H_2SO_4), redoxolysis (Fe^{3+})	45°C	1.2 - 2.0	Batch reactor	1%	**Cu** 89% (76 mg/g), **Ni** 81% (16.2 mg/g), **Zn** 83% (66.4 mg/g)	Ilyas et al., 2007
Acidithiobacillus sp., Gallionella sp., Leptospirillum sp.	redoxolysis (Fe^{3+})	30°C	1.5 - 2.5	Batch reactor	2%	**Cu** 95% (219 mg/g)	Xiang et al., 2010

Microorganism(s)	Leaching mechanism	Temp.°C	pH	Reactor type	Pulp density (w/v)	Leached metals (percentage, mg/g PCB)	References
Chromobacterium violaceum, Pseudomonas fluorescens, Pseudomonas plecoglossicida	Complexolysis (CN^-)	30°C	7.2 - 9.2	Batch reactor	various	**Au** 69% (not specified)	Brandl et al., 2008
Acidophilic consortium (genera Acidithiobacillus and Gallionella)	Redoxolysis (Fe^{3+})	30°C	2.0	Batch reactor	1.2%	**Cu** 97% (626 mg/g), **Al** 88% (34 mg/g), **Zn** 92% (28 mg/g)	Zhu et al., 2011
A. ferrooxidans, A. thiooxidans	Redoxolysis (Fe^{3+})	28°C	1.8 - 2.5	Batch reactor	0.8 - 1.9%	**Cu** 99% (90 mg/g)	Wang et al., 2009
A. ferrooxidans, Leptospirillum ferrooxidans, A. thiooxidans	Acidolysis (H_2SO_4), redoxolysis (Fe^{3+})	25°C	1.7	Batch reactor	1%	**Cu** 95% (106 mg/g)	Bas et al., 2013
A. ferrooxidans, A. thiooxidans	Acidolysis (H_2SO_4), redoxolysis (Fe^{3+})	28°C	1.5 - 3.5	Batch reactor	3% (increased gradually)	**Cu** (94%), **Ni** (89%), **Zn** (90%)	Liang et al., 2010

Microorganism(s)	Leaching mechanism	Temp.°C	pH	Reactor type	Pulp density (w/v)	Leached metals (percentage, mg/g PCB)	References
Chromobacterium violaceum, Pseudomonas aeruginosa	Complexolysis (CN^-)	30°C	9.2	Batch reactor	1%	**Cu** 83% (105 mg/g), **Zn** 49% (27 mg/g), **Au** 73% (0.01 mg/g), **Ag** 8% (0.03 mg/g)	Pradhan and Kumar, 2012
Pseudomonas chlororaphis	Complexolysis (CN^-)	25°C	7.0	Batch reactor	1.6%	**Au** (8%), **Ag** (12%), **Cu** (52%)	Ruan et al., 2014
A. ferrooxidans, A. thiooxidans, Thiobacillus denitrificans, Thiobacillus thioparus, Bacillus subtilis, Bacillus cereus	Acidolysis (H_2SO_4), redoxolysis (Fe^{3+}), complexolysis (surfactants)	22 - 37°C	5.0 - 7.0	Batch reactor	1%	**Cu** 53% (22 mg/g), **Ni** 48.5% (6.4 mg/g), **Zn** 48% (6 mg/g)	Karwowska et al., 2014
A. caldus, Leptospirillum ferriphilum, Sulfobacillus benefaciens, Ferroplasma acidiphilum	Acidolysis (H_2SO_4), redoxolysis (Fe^{3+})	37°C	1.7	Batch reactor	1%	**Cu** 99% (29 mg/g)	Bryan et al., 2015
A. ferrivorans, A. thiooxidans	Acidolysis (H_2SO_4), redoxolysis (Fe^{3+})	23°C	1.0 - 2.0	Batch reactor	1%	**Cu** 98% (164 mg/g)	Işıldar et al., 2015

Microorganism(s)	Leaching mechanism	Temp.°C	pH	Reactor type	Pulp density (w/v)	Leached metals (percentage, mg/g PCB)	References
Sulfobacillus thermosulfidooxidans, Thermoplasma acidophilum	Acidolysis (H_2SO_4), redoxolysis (Fe^{3+})	45°C	1.5 - 2.7	Column reactor	n/a (10 kg)	**Cu** 86% (76 mg/g), **Zn** 80% (71 mg/g), **Ni** 74% (15 mg/g), **Al** 64% (6.5 mg/g)	Ilyas et al., 2010
A.ferrooxidans, A. thiooxidans	Acidolysis (H_2SO_4), redoxolysis (Fe^{3+})	28°C	1.1 - 1.6	CSTR	1%	**Cu** 99% (151 mg/g)	Mäkinen et al., 2015
S. thermosulfidooxidans	Acidolysis (H_2SO_4), redoxolysis (Fe^{3+})	45°C	2.0	CSTR	1.5 - 3.5%	**Cu** 95% (105 mg/g), **Al** 91% (19 mg/g), **Zn** 96% (18 mg/g), **Ni** 94% (18 mg/g)	Ilyas and Lee, 2014b
A. ferrooxidans	Redoxolysis (Fe^{3+})	30°C	2.0	Column + CSTR	0.4 - 1.6%	**Cu** 92 (582 mg/g), **Zn,** 90% (37 mg/g) **Al,** and 59% (41 mg/g)	Nie et al., 2015
A. ferrooxidans	Redoxolysis (Fe^{3+})	30°C	2.0	Batch reactor	1.2%	**Cu** 96% (604 mg/g)	Nie et al., 2014

3.3 Biorecovery of metals

Many well-practiced conventional techniques such as solvent extraction, cementation, ion exchange, precipitation, adsorption, and electrowinning, enable selective recovery of metals from leachate solutions. They are used to selective recovery of As, Cd, Se, Cu, Fe, Ni, Zn, Cr, Pb from waste waters (Fu and Wang, 2011), industrial waste water (González-Muñoz et al., 2006) and solid wastes (Cui and Zhang, 2008; Tuncuk et al., 2012). There is also an increasing interest on biosorbents for precious metal recovery from aqueous solutions (Das, 2010). The selection of an appropriate technique for metal recovery depends on process parameters, such as metal concentration in leachate liquor and metal-selective behavior of the selected technique. At industrial scale, many recovery processes are a combination of the above-mentioned conventional techniques (Schlesinger et al., 2011). Several biological processes, such as biosorption, bioreduction, biomineralization, bioprecipitation could be alternative metal recovery methods (Hennebel et al., 2015).

Currently, research on selective recovery of metals from electronic waste leachate liquors is limited. Compared to primary ores, electronic waste materials are very concentrated in metals and complex owing to prevalence of a large number of metals (Ongondo et al., 2015). Usage of biomass-based techniques for the recovery of metals is an emerging field with vast potential. Biorecovery of metals from leachate solution is shown to be effective, in particular with relatively low metal concentrations (Gadd, 2010). Biorecovery of metals from electronic waste could provide a viable, environmentally friendly options.

Several mechanisms of cation uptake by the cells are proposed, such as (i) binding on cell surfaces, (ii) resistance/detoxification mechanisms, (iii) bioaccumulation within the cell wall, (iv) active translocation inside the cell through metal binding proteins, as well as mineralization actions such as (v) interaction with extracellular polymers, or (vi) volatilization through enzymes (Das, 2010; Andrès and Gérente, 2011). In this section, biosorption, bioreduction, biomineralization and bioprecipitation from aqueous solutions are given.

3.3.1 Biosorption

Biosorption is a feature of microbial biomass to bind and concentrate metals from aqueous solutions. All prokaryotic cells, e.g. bacteria; and eukaryotic cells of the phyla algae, fungi, and yeasts bind metals (Das, 2010). It is typically carried out by inactive biomass. Common biosorbents include various compounds with relatively high surface amine functional group content. This is generally due to the ability of the positively charged amine groups to attract metal ions (Mack et al., 2007). Recently, Tanaka and Watanabe (2015) showed that Pt absorbed on bacterial cells has a fourfold coordination of chlorine ions, similar to Pt_4^{2-}, which indicated that sorption occurs on the protonated amine groups of the bacterial cells.

The biosorption rate is related to the structure of the microbial cell wall, the cation chemistry, the medium conditions, nature of functional groups and surface area (Andrès and Gérente, 2011; Ilyas and Lee, 2014a). On a cellular scale, biosorption takes place at the cell wall or by various metabolites, e.g. metal-binding peptides, polysaccharides, extracellular polymeric substances (EPS) (Gadd, 2010). The mode of solute uptake by inactive cells is extracellular, where the chemical functional groups of the cell wall play vital roles in biosorption (Ilyas and Lee, 2014a). Several functional groups are present on the cell wall including carboxyl, phosphoryl, amine and hydroxyl groups (Wang and Chen, 2009). Carboxylic groups of the cell wall peptidoglycan of the Actinobacteria *Streptomyces pilosus* are responsible for the binding of divalent metal ions (Tunca et al., 2007).

Algae, fungi, yeasts, and bacteria play a role as biosorbent for precious metals (Mack et al., 2007). An overview of sorption of precious metals is given below in Table 3-3. Active cells of the green alga *Chlorella vulgaris* has a high efficiency in removing Au from solution (Ting and Mittal, 2002). The brown alga *Sargassum natans* is highly selective towards Au (Das, 2010). Inactivated cells of the related species *Fucus vesiculosus* can recover elemental Au as nanoparticles (Mata et al., 2009). The fungal cells of *Aspergillus niger*, *Mucor rouxii* and *Rhizopus arrihus* were found to take up gold along with other precious metals (Syed, 2012). Two strains of a fungus, *Cladosporium cladosporioides* showed preferential sorption of Au (Pethkar et al., 2001). Among the gram-negative bacteria, *Acinetobacter calcoaceticus*, *Erwinia herbicola*, *Pseudomonas aeruginosa and Stenotrophomonas maltophilia* are shown to be capable of Au biosorption (Das, 2010; Ye et al., 2013). It is a viable alternative for metals recovery from dilute solutions (Mack et al., 2007; Wang and Chen, 2009).

Table 3-3: Biosorption of precious metals from aqueous solutions by algal, bacterial, and fungal cells.

Metal	Biosorbent	Mode of action	Uptake (mg/g biomass)	Temp°C	pH	References
			Algae			
Au(III)	*Sargassum natans*	Inactive biomass	82.7	ambient	7.0	Kuyucak and Volesky, 1988
Au(III)	*Chlorella vulgaris*	Inactive cells	98.5	ambient	2.0	Darnall et al., 1986
Au(III)	*Turbinaria conoide*	Inactive cells	34.5	25°C	2.0	Vijayaraghavan et al., 2011
Au(Cl$_4^-$)	*Fucus vesiculosus*	Inactive cells	75	ambient	7.0	Mata et al., 2009
Au(CN$_2^-$)	*Bacillus subtilis*	Inactive biomass	92.5	ambient	2.0	Niu and Volesky, 2000
Ag(II)	*Bacillus cereus*	Inactive biomass	91.4	30°C	4.0	Li et al., 2011
			Bacteria			
Au(III)	*Escherichia coli*	Active cells	115	ambient	6.5	Deplanche and Macaskie, 2008
Au(CN$_2^-$)	*Corynebacterium glutamicum*	Inactive cells	421.1	25°C	5.5	Park et al., 2012
Pd(II)	*Delsufovibrio desulfuricans*	Active cells	190.0	37°C	3.0	de Vargas et al., 2004
Pd(II)	*Desulfovibrio fructosivorans*	Active cells	63.8	37°C	2.3	Mikheenko et al., 2008

Metal	Biosorbent	Mode of action	Uptake (mg/g biomass)	Temp°C	pH	References
Pd(II)	*Escherichia coli*	Inactive cells	265.3	25°C	3.0	Park et al., 2010
Pd(II)	*Corynebacterium glutamicum*	Active cells	176.8	25°C	2.0	Won et al., 2011
Pd(II)	*Escherichia coli*	Inactive biomass	265.3	25°C	3.0	Won et al., 2010
Pt(II)	*Escherichia coli*	Inactive biomass	108.8	25°C	acidic	Won et al., 2010
Pt(II)	*Bacillus subtilis*	Active cells	100.0	25°C	2.0	Tanaka and Watanabe, 2015
Pt(IV)	*Shewanella putrefaciens*	Active cells	100.0	25°C	4.0	Tanaka and Watanabe, 2015
Pt(IV)	*Delsufovibrio desulfuricans*	Active cells	90.0	37°C	3.0	de Vargas et al., 2004

Fungi

Metal	Biosorbent	Mode of action	Uptake (mg/g biomass)	Temp°C	pH	References
Au(III)	*Fomitopsis carnea*	Inactive cells	94.3	25°C	8.0	Khoo and Ting, 2001
Au(III)	*Cladosporium cladosporioides*	Biomass beads	101.0	ambient	4.0	Pethkar et al., 2001
Au(III)	*Aspergillus niger*	Active cells	197.0	ambient	7.0	Kuyucak and Volesky, 1988
Pt(IV)	*Saccharomyces cerevisiae*	Inactive cells	44.0	ambient	3.5	Xie et al., 2003

3.3.2 **Reductive bioprecipitation**

Reductive bioprecipitation describes enzymatically assisted metal precipitation from a positive valence to a zero-valent state (Rawlings et al., 2003). Bioreduction of metals takes place either by direct contact to the cell surface of or through extracellular electron shuttles (Manzella et al., 2013). Biomineral formation takes place through a number of mechanisms, e.g. bioprecipitation, intracellular accumulation, nanoparticle formation, bioreduction, redox immobilization (Gadd, 2010). Some secondary metabolites, namely polyketides and non-ribosomal peptides, are produced by the bacteria to promote environmental fitness and bind metals (Johnston et al., 2013). Polyketides are a class of secondary metabolites giving organisms some survival advantage. Many mycotoxins produced by fungi are polyketides. Although involvement of multiple electron transfer steps is known in metal bioreduction, the exact mechanism remains unclear.

Enzymatic mechanisms promote metal ion reduction under favorable conditions, independently of cell metabolism. The identification of the enzymatic mechanisms may indicate their prevalence in growth-decoupled action. In some occasions, metal bioreduction takes place via direct electron transport reactions producing crystals of metal oxide or base metal accumulated on the cell surface (Deplanche et al., 2005). Bacterial biofilms are present on the surface of Au nuggets (Reith et al., 2009), despite the inherently toxic characteristics of soluble Au, the bacterial cells are speculated to accumulate Au intracellularly (Johnston et al., 2013). A number of hyperthermophilic and mesophilic dissimilatory iron-reducing bacteria and archaea are shown to be capable of producing precipitates of elemental gold from Au^{3+} cations (Kashefi et al., 2001). Bioreduction of Au^{3+} is an enzymatically catalyzed reaction, and dependent on electron donor supply, e.g. gaseous hydrogen. Au precipitated extracellularly with much of the elemental Au attached to the outer surface of the cells. The mechanism of reductive precipitation of Au by ferric iron-reducing microorganisms observed to be significantly different from the bioaccumulation of Au (Kashefi et al., 2001).

Selective reductive bioprecipitation of metals from electronic waste leach liquor using *Desulfovibrio desulfuricans* biomass has been investigated in batch tests by Creamer et al., (2006). It proved effective for Au^{3+}, Pd^{2+}, and Cu^{2+} in a three-step process. In the first step active cells selectively precipitated Au^{3+} to elemental Au from leach liquor, while Pd^{2+} precipitation was inhibited due the presence of high amount of Cu^{2+}. In the second step, the pretreated (palladised) biomass was used to catalyze the conversion of Pd^{2+} as elemental Pd^{0}.

In the third step, the remaining leachate solution was treated by biogas generated by *Klebsiela pneumoniae* or *Escherichia coli*, where Cu is removed as a mixture of hydroxide and sulfate salts. In the Au and Pd recovery steps, hydrogen sparing enabled the initiation of metal reduction (Creamer et al., 2006).

3.3.3 Biomineralization

Investigations on *Cupriavidus metallidurans* have revealed that it bioaccumulates inert Au nanoparticles within its cytoplasm as a mechanism to protect itself from soluble Au (Reith et al., 2009). Gold-resident bacterium *Delftia acidovorans* produces a metabolite, namely delftibactin, that assists its survival (Johnston et al., 2013). It is speculated this bacterium secretes metabolite against toxicity, which in turn enable biomineralization of elemental Au.

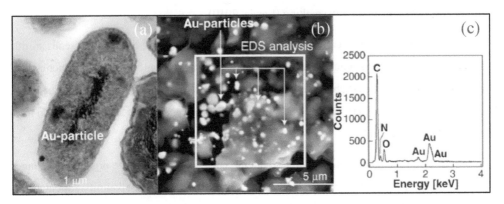

Figure 3-5: Gold biomineralization in the bacterium Cupriavidus metallidurans; Transmission electron micrograph (TEM) of ultra-thin section containing Au nanoparticle (A), Scanning electron microscopy micrograph (SEM) (B), with energy dispersive X-ray analysis (EDS) (C) (Source: Reith et al., 2009).

Foulkes et al. (2016) demonstrated the presence of a novel mechanism responsible for the biomineralization of Pd(II) in aerobically grown cultures of *E. coli*, catalyzed mainly by molybdenum-containing enzyme systems. The strain that lacked all the enzymes did still reduce the palladium, although this took 7 h, compared with less than 30 min by the wild-type strains. Biomineralization still occurs in strains without these enzymes albeit at a much lower rate. The real-time analysis showed that bioPd(0) were mineralized outside the cells.

Biomineralization of platinum group metals (PGMs), the main metals of interest in WEEE to be recovered, is an emerging field of biotechnology. It is a viable alternative for the selective recovery of metals from complex WEEE leachates, containing these precious metals. Moreover, the microbial cells are selective towards individual metals, which gives this technology advantages over conventional technologies.

3.3.4 Biogenic sulfide precipitation of metals

Sulfide precipitation of metals is an established technique particularly for wastewaters with high metal concentrations, e.g. acid mine drainage. Biotechnological applications of bacterial sulfate reduction have a couple of advantages such as cost-effectiveness and lower volume of sludge generation as compared to hydroxide precipitation. Moreover, it enables selective precipitation of metal sulfides in a pH dependent stoichiometry (Sahinkaya et al., 2009; Sampaio et al., 2010). It attract scientific interest due to its advantages such as lower solubility of precipitates, potential for selective metal removal, fast reaction rates, and potential for re-use of metal sulfide precipitates by smelting (Lewis, 2010).

Sulfide precipitation can be carried out using either aqueous (Na_2S, $NaHS$) or gaseous sulfide sources (H_2S). Many studies investigated the removal efficiency, reaction kinetics and crystallization properties (Sahinkaya et al., 2009; Mokone et al., 2010; Janyasuthiwong et al., 2015). Solubility characteristics of metal sulfide compounds enable selective precipitation from a mixed metal solution.

In biogenic sulfidic precipitation, sulfate-reducing bacteria (SRB) oxidize organic compounds by sulfate as an electron acceptor and generate sulfide (S^{2-}) and alkalinity. Spontaneous reduction of SO_4^{2-} under ambient conditions occurs solely in the presence of microorganisms. In a continuous system (SRB) produce gaseous hydrogen sulfide in the first reactor and the subsequent metal precipitation takes place in the second reactor (Jong and Parry, 2003). This process is based on the ability of SRB to reduce sulfates to sulfides, which form insoluble precipitates of metal sulfides (Lewis, 2010). Generic reactions (3-9) and (3-10) are given below.

$$\text{Organic matter (C, H, O)} + SO_4^{2-} \rightarrow HS^- + HCO_3^- \qquad (3\text{-}9)$$

$$Me^{2+} + HS^- \rightarrow MeS \downarrow + H^+ \qquad (3\text{-}10)$$

Cao et al. (2009) investigated the precipitation characteristics of metal cations from WEEE leachate solution. Metal concentrations of 20, 5, 2, and 0.5 g/L for Mg, Fe, Ni, and Cu, respectively were used by for sulfidic precipitation with biologically-produced H_2S gas. The sulfide concentration is found to strongly influence the precipitation efficiency. The order of the removal rate of Cu as the best, followed respectively by Fe, Ni, and Mg. Cu removal was 100% in two experiments with various sulfide concentrations, while Fe removal was 62.7–100% and Ni removal was 46.4–100%. Moreover, it was found that the efficiency of metal precipitation with biogenic H_2S depended on the type of the reactor type. The pH had a significant influence on the metals removal since the rate of H_2S dissolution is faster at high pH values. pH of the bioleaching solution had no influence on the precipitation efficiency, provided that the H_2S concentration was sufficient for metal sulfide precipitation (Cao et al., 2009).

Table 3-4: Overview of metal biorecovery techniques.

Method	Targeted metals	References
Biosorption	Precious metals; Ag, Au, Pd, Pt	Vijayaraghavan et al., (2011); Park et al., (2010); Mikheenko et al., (2008); Xie et al., (2003)
Reductive bioprecipitation	Precious metals; Au, Pd, Pt	Kashefi et al., (2001); Creamer et al., (2006)
Biomineralization	Cu and precious metals; Au, Pd	Reith et al., (2009); Johnston et al., (2013)
Bioprecipitation (sulfide)	Base metals: Cu, Ni, Zn	Cao et al., (2009); Sahinkaya et al., (2009); Sampaio et al., (2009); Janyasuthiwong et al., (2015)

3.4 Conclusions

Electronic waste, in particular discarded PCB, is a promising secondary source of metals. Cu is the predominant metal by weight, along with substantial amount of other base metals and precious metals. Biotechnological metal recovery techniques enable more environmental friendly and cost-effective processes and are expected to play a significant role for sustainable development. Research work on metal recovery from electronic waste has focused on acidophilic and cyanogenic bioleaching processes.

Currently, very high (>99%) Cu bioleaching efficiency has been achieved in laboratory scale batch and continuous setups. The main influencing operating parameters were pH, redox potential, microbial activity, and pulp density. Moreover, recent studies showed that the leaching process could be significantly accelerated, the metal leaching efficiency could be improved and that bacteria can process even higher loads of waste in reactor setups. Cyanogenic bioleaching, on the other hand, is a field of research at its infancy. Many microbes are shown to decompose the excess biogenic cyanide which might have implication in a scaled-up system.

Biorecovery of metals from leachate solution is still at an early stage of research. The composition of leachate solutions from waste materials is very complex, which requires novel strategies to recover metals from. There is need for research on selective recovery of metals from leachate solution. In this direction, multi-step approaches might be considered. Biotechnology will play an increasingly significant role in the future of mineral and waste processing. Recent studies showed that bioleaching of Cu, the primary metal of PCB, is feasible in continuous reactor setups. Ongoing investigations aim to provide further insights into mechanisms of bioprocessing of waste material, improvements in the efficiency and expansion of the range of metals amenable for recovery. Finally, many studies focus on optimizing the process efficiency in terms of duration and economic feasibility. Biotechnologies have a vast potential in treatment of waste for metal recovery. Should the complex bioleaching processes be optimized and the operational parameters are improved, biotechnologies can be play a prominent role in biorecovery of metals from WEEE.

Chapter 4.

Characterization of discarded printed circuit boards and a multi-criteria analysis approach for metal recovery technology selection

Abstract

Waste electrical and electronic equipment (WEEE) is an important secondary source of metals. Particularly discarded printed circuit boards (PCB), despite their relative small weight fraction, are a valuable secondary source of copper (Cu), iron (Fe), aluminum (Al), nickel (Ni), zinc (Zn), gold (Au), silver (Ag) and palladium (Pd). The concentration of these metals in the discarded PCB varies greatly, depending on the source of the material, manufacture year, type of the board and the manufacture technology. There is not yet a standard method to characterize and assay the metals in discarded PCB. In this chapter, a new characterization and total metal method is presented, and its application on a number of PCB from desktop computers, laptop computers, computer parts, mobile phones and telecom devices is carried out. Furthermore, multi-criteria analysis (MCA) using analytic hierarchy process (AHP) is used to rank and select the most appropriate technology for metal recovery from discarded PCB. The analysis is based on a number of criteria, viz. economic, environmental, social and technical criteria, along with a number of sub-criteria. The metal concentration assayed resulted in a varied concentration of Cu (9.3% – 38.4% by weight) across particle sizes (<500 µm, 500 – 1600 µm and 1600 – 2500 µm) and board types, along with considerable concentrations (by weight) of Fe (1.4% – 7.5%), Al (0.1% – 3.0%), Ni (0.1% – 2.4%), Zn (0.03% – 1.0%), Pb (0.04% – 0.9%) and Au (21-320 ppm). The MCA study revealed that biohydrometallurgical route is slightly preferred over a hydrometallurgical routes in technology selection for sustainable metals recovery from WEEE.

4.1 Introduction

In the global environmental agenda, waste in no longer regarded as an unwanted material with negative value. It is rather a material with potential for resource recovery, and a secondary source of raw materials. Global waste generation almost doubled from 0.64 kg to 1.2 kg/person per day between 2005 and 2012, and it is likely to reach 2.1 kg by 2025 (Karak et al., 2012). This signifies an enormous potential for resource recovery as discarded materials play a crucial role in transition to a circular economy (Jones et al., 2013). In integrated waste management, the trend has shifted from systematic and controlled disposal of waste in designated areas to resource recovery and zero waste (Mueller et al., 2015). In order to facilitate the shift to a circular zero waste economy, several material recovery strategies with minimized dissipative losses are envisioned to be developed.

Extensive research efforts enabled many novel resource recovery technologies from various types of waste. Most widely reported are energy and nutrient recovery from domestic wastewater (Batstone et al., 2015), energy recovery from organic solids in waste-to-energy (WtE) plants (Cucchiella et al., 2014), and increasingly metal recovery from industrial (Jadhav and Hocheng, 2012) and electronic waste (Akcil et al., 2015). Modern electronic devices contain almost all the metals of the periodic table at varying concentrations (Graedel and Erdmann, 2012), and make excellent candidates of secondary resources of critical metals. The wide variety of available technologies enable engineers and researchers to design the best system for resource recovery from waste materials. There is no 'one-fits-all technology' for the recovery of valuable resources from increasingly complex anthropogenic waste streams. Understanding the composition and characteristics of the waste materials is important to estimate their reuse and recyclability potential (Ahluwalia and Nema, 2007).

Waste electrical and electronic equipment (WEEE) makes up to 8% of the total municipal waste in developed economies with an increasing generation rate (Robinson, 2009). Despite hazards stemming from its improper management, WEEE is also an important secondary source of metals. Distinctive to primary ores, the abundance of the metals, their complexity degree, and chemical structure are very variable (Ongondo et al., 2015). Typically, a discarded printed circuit board (PCB), an integral part of every electronic device, includes high concentration of metals, between (in %, w/w) $10 - 35$ copper (Cu), $1 - 12$ iron (Fe), $0.5 - 5$ nickel (Ni), $1 - 15$ aluminum (Al), along with gold (Au) (in ppm) $50 - 500$ (Yamane et al., 2011; Oguchi et al., 2012; Hadi et al., 2015). These metals are mostly found in complex alloys in their elemental metallic forms (Tuncuk et al., 2012). This urges novel recovery strategies so as to sustainably and efficiently recover metals from end-of-life (EoL) devices.

Technology selection prior to applied research activities is an essential step in research development and innovation (RD&I) processes. A large fraction of the total research life cycle costs and resource allocation are determined by decisions made before the initialization of research activities (Georgiadis et al., 2013). Resource recovery from urban waste streams (Urban mining) is a complicated process that involves multiple specialized criteria (Soltani et al., 2015). Many resource recovery options are already available or currently under development, and the technology selection involves decision-making based on the characterization of the waste material, available methods, technological readiness level (TRL), techno-economic assessment, and environmental profile (Xia et al., 2015). In method selection, the suitability, validity and user-friendliness of the methods are the important factors to be considered. Often, multiple criteria are involved in decision-making for technology selection (Mateo, 2012). Joint consideration of the heterogeneous and uncertain information demands a systematic and understandable structure to interpret the technical information (Huang et al., 2011).

Multi-criteria analysis (MCA) is a tool to analyze the viable technology options and select the most appropriate for the desired objective. Multi-criteria decision analysis (MCDA) emerged as a formal methodology to analyze available technical information and stakeholder values in environmental decision making (Khalili and Duecker, 2013). MCA facilitates the aggregation of the each selected criterion, techno-economic expedience and environmental profile with the relevant criteria of cost-effectiveness, technological status, socially-responsible operation and sustainable development (Sliogeriene et al., 2013). There is a significant growth in environmental applications of MCA over the last decade, from a total number of 26 to 200 between 2000 and 2010, across all environmental application areas (Huang et al., 2011).

The Analytic hierarchy process (AHP) method is increasingly used for waste management projects, which requires a joint evaluation of technical, economic, environmental and social aspects. AHP is based on the pairwise comparison principle that is used for deriving weights of importance of the criteria and relative rankings of alternatives for each criterion. Quantitative and qualitative factors are weighted by computing numerical comparison and reciprocal scales (Pohekar and Ramachandran, 2004). Between 1990 and 2010, out of 312 MCDA papers, 150 used the AHP methodology (Huang et al., 2011). A further breakdown (methodology used/total projects/percentage distribution) showed that waste management (15/30, 50%), natural resource management (7/14, 50%), and sustainable engineering (18/28, 64%) are topics in which MCA is widely applied.

Current WEEE recycling practices include physical/mechanical, thermal treatment, hydrometallurgical or biohydrometallurgy routes, or commonly a combination of them (Chapter 3). Typically, physical pretreatment is followed by a metals extraction process (Tuncuk et al., 2012). In this step, conventional chemical methods or emerging biological routes, and a hybrid mixture of the two can be applied (Ilyas et al., 2015). In this work, the main objectives are **(1)** to develop and apply a characterization method to a complex, polymetallic waste material, and **(2)** to select an appropriate metal extraction technology from discarded PCB. In this chapter, a method is described to characterize discarded PCB. Furthermore, the technological suitability of the metal recovery alternative was evaluated using the AHP methodology by the selection of significant criteria and relevant technology selection indicators.

4.2 Methodology

4.2.1 Source and preparation of the materials

Discarded electronic devices, namely desktop computers, laptop computers, mobile phones, and telecom devices were collected from the IT department of UNESCO-IHE (Delft, the Netherlands) and SIMS recycling company (Eindhoven, the Netherlands). The discarded devices were manually dismantled and the PCB were taken out. The PCB of various devices were grouped into 7 categories, namely **(1)** desktop computer boards, **(2)** computer parts, **(3)** desktop computer boards without components, **(4)** laptop computer boards, **(5)** mobile phone boards, **(6)** telecom boards 1 and **(7)** telecom boards 2. Batteries, heat sinks and microprocessors were removed from the PCB of the desktop computers. Parts of the desktop computers, i.e. video cards and sound cards, were manually removed in order to determine their metal concentrations. Components of the boards such as connectors, capacitors, and integrated chips were removed using specialized hardware equipment (screw drivers for electronics) and a thermal gun (Gamma, HG2000E, the Netherlands). The PCB were washed with distilled water, cut manually in pieces of approximately 5 cm × 5 cm, crushed by a cutting mill (Retsch 2000, Germany) using a 2.5 mm sized trapezoid sieve. Ground PCB materials were washed with distilled water and dried overnight prior to metal analysis. For particle size distribution analysis, the samples were sieved manually (Fritsch, UK) to a particle sizes **(1)** smaller than 500 µm, **(2)** between 500 – 1600 µm, and **(3)** between 1600 – 2500 µm. Screen analysis was performed by passing 100 g of the crushed material in a mechanically shaking sieve apparatus (ROTAP, W.S. Tyler, USA). Loss on comminution (LoC) was analyzed by weighting each material group before crushing and after sieving.

4.2.2 Waste characterization and total metal assay

A modified EPA method (3052-1996: Microwave assisted acid digestion of siliceous and organically based matrices) was used for the characterization of the waste material. The fluorocarbon polymer reaction vessels were acid-washed in 6 M HCl overnight before each digestion procedure. 0.5 g of the crushed sample was added to 12 mL of nitro-hydrochloric acid, a mixture of concentrated HCl and NHO_3 at a ratio of 3:1 (v/v), respectively, in a laboratory microwave accelerated reaction unit (CEM Duo Temp Mars 5, USA). The analytes were placed in the vessels and left for 1 h for their stabilization and the emission of excess gases. The vessels were then capped, and placed in the microwave system for digestion. The mixture containing acid and the sample was gradually heated up to 175°C in 5.5 min and remained at 175°C for 9.5 min (CEM, MarsXpress, USA). At the end of the program, the vessels were cooled to room temperature in a fume hood and the pressure inside the vessels was released slowly. The digestates were filtered (Whatmann, G/C) twice, and centrifuged for 20 min at 2,500 rpm and membrane filtered (0.22 μm) to remove any remaining undissolved solid particles that might block the nebulizer of the inductively couple plasma mass spectrophotometer (ICP–MS). The analytes were serially diluted, and analyzed for their metal concentrations. Modifications to the US EPA method 3052 included: **(1)** increasing the ramping and total duration of microwave digestion, **(2)** varying the ratio of sample mass to acid mixture volume and **(3)** alterations to the composition of the acid digestion mixture.

4.2.3 Analytical methods

Metal analyses were performed by inductively coupled plasma mass spectrophotometer (ICP-MS) (XSERIES 2, Thermo Scientific, USA) and optical emission spectrometer (ICP-OES) (Perkin Elmer, Optima 8300, USA). Periodic metal measurements were carried out by atomic absorption spectrophotometer in flame mode (AAS-F) (Varian 200, USA). The readings for the following wavelengths were recorded for individual metals (in nm), Al 396.153, Ag 328.068, Cd 226.502, Co 228.616, Cr 267.716, Cu 324.752, Zn 206.2, Fe 238.204, Ni 231.604, Pb 220.353, Nd 401.2, La 379.478, Ce 418.6, Au 267.595, Pt 265.945, Pd 363.47 in ICP-OES.

The materials were examined using optical microscopy (BX63, Olympus, Japan), binocular microscopy (SZ61, Olympus, Japan), scanning electron microscopy (SEM) equipped with energy dispersive X-ray analyzer (EDS) (JSM-6010LA, JEOL, Japan) and X-ray powder diffraction (Bruker B8 ADVANCE, USA). The intrinsic value of each metal was calculated according to their mass fraction and their price on the international market values at the London Metal Exchange on a 3-month average (LME, 2016). The average concentration values were calculated according to their weighted fraction in respective particle sizes.

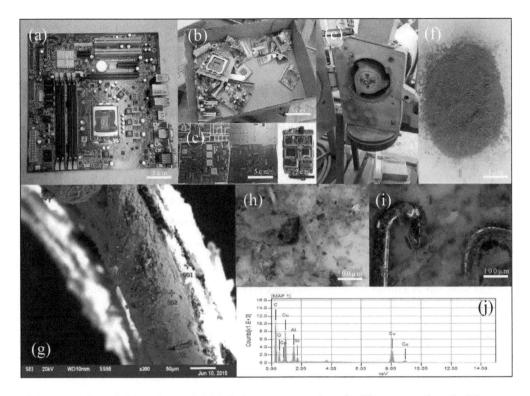

Figure 4-1: Discarded board material (a) desktop computer boards, (b) computer boards, (c) telecom boards, (d) cell phone boards before crushing, (e) Retsch SM2000 crusher, (f) crushed and sieved boards material (<500 μm), (g) SEM micrographs of crushed board, (h) and (i) binocular microscope pictures, (j) EDS diagram of the crushed boards.

4.2.4 Multi-criteria analysis for technology selection

The main steps of MCA are as follows: **(1) Framing the MCA approach**, discerning the alternatives, establishing the criteria, defining the goal, and any conflicting points, and the degree of uncertainty, **(2) Assigning the criteria weights,** defining the weights, that show the relative importance of criteria in the multi-criteria problem under consideration and **(3) Construction of the evaluation matrix,** the matrix model constituted a process from which the essence of the goal was extracted. At the end of this step, the MCA goal presented the best available alternative.

The major economic driver for recycling of WEEE is the recovery of valuable metals, thus potential revenue from the recovery of metals is a major incentive. Therefore, economic criteria are weighted highest among the considered factors. However, several other criteria are to be taken into account. Environmental, social and technical criteria are very significant in selection of a technology in line with global sustainable development goals. Metal recovery from WEEE is traditionally carried out by pyrometallurgical and hydrometallurgical methods, each with their own advantages, limitations, and drawbacks (Cui and Zhang, 2008). Biohydrometallurgy, using microbes for metal recovery, enables more environmental-friendly and cleaner processes (Hennebel et al., 2015). Table 4-1 overviews the considered metal recovery technology options.

4.2.4.1 *Metal recovery technology alternatives*

The main goal is to make a selection between a conventional (hydrometallurgical) and an emerging (biohydrometallurgical) method for metal recovery technology development. Thus, hydrometallurgical and biohydrometallurgical metal extraction routes are considered as the alternatives in the MCA system. Pyrometallurgy is not considered as an alternative in the present technology selection, as it is already at an industrial scale and a number of industrial plants readily operate to recover valuable metals from discarded PCB (Cui and Zhang, 2008). The selection of a novel recovery technology is considered to be selected among the emerging technologies.

There are two alternatives, namely:

A_1: Metal recovery from WEEE by using **hydrometallurgy**, through conventional chemically mediated metallurgical processes.

A_2: Metal recovery from WEEE by using **biohydrometallurgy**, through biologically catalyzed processes.

In the first alternative, A_1, the metals are extracted through a hydrometallurgical route, i.e. leaching of metal(s) of interest in corrosive oxidative acidic and/or alkaline conditions to solubilize the metals into a leachate solution. Many studies investigated the use of conventional or innovative chemical reagents to recovery metals from a polymetallic anthropogenic secondary sources, with a number of pilot scale demonstration already available (Tuncuk et al., 2012; Hadi et al., 2015). In the second alternative, A_2, metal recovery from WEEE is carried out by using an emerging technology: biohydrometallurgy. The principles of this technology lies at using microbes to extract metals in natural processes and is widely used in the production of metals form primary ores (Vera et al., 2013). Globally, more than 15% of copper and 5% of gold is produced by using bacteria from their primary ores (Johnson, 2014). Recently, many researchers reported promising by biohydrometallurgical treatment and metal recovery from secondary resources (Jadhav and Hocheng, 2012).

Table 4-1: Metal recovery approaches from electronic waste.

Parameters	Pyrometallurgy	Hydrometallurgy	Biohydrometallurgy
Environmental impact	High; due to hazardous emissions and high energy usage	Moderate; due to usage of toxic chemicals and water consumption	Low; no emissions, water can be recirculated
Economic	Capital intensive, low job creation	Low first investment costs, high operating cost	Low investment and operating costs
Social acceptance	Low, infamous track record of the mining industry	Medium, some toxic reagents and end products	High, cleaner processes and self-control of the pollution
Recovery rate	Low, only a fraction of metals	High recovery efficiency	Medium efficiency, high selectivity
Final residue	High; solid slags, and gaseous emissions	Low; non-metallic part of the waste and waste water	Low; non-metallic part of the waste and very low waste water
Process conditions	Harsh conditions due to furnaces and smelting	Harsh conditions, a few toxic chemicals	Safe conditions, low-to-none toxic chemicals
Level of advancement	High, established full scale technology	Medium, many pilot scale demonstrations	Low TRL, no pilot scale study
Advancement requirement	Abatement of environmental impacts	Selectivity towards individual metals, scale-up studies	Fundamental research, scale-up studies
Feasible applications	Low, only high grade WEEE	High, all metals and their alloys	Medium, restrictions due to toxicity on bacteria

4.2.4.2 *Selection of the criteria and establishing criteria hierarchy*

Selection of relevant criteria requires the evaluation of different options considering all sustainability dimensions (Serna et al., 2016). Evaluation criteria have been selected to meet the assumptions of a sustainable development concept and take into account the economic, environmental, social and technical aspects of the analyzed option. The selected criteria are analogous and introduce the assessment of varying sustainability dimensions in the context of technology development.

Figure 4-2 shows the selected criteria and their hierarchy. The selection was based on the key parameters reported above in Table 4-1 as well as the data reported in the literature (Cui and Zhang, 2008; Tuncuk et al., 2012; Marques, 2013; Cucchiella et al., 2015; Hadi et al., 2015). The selected criteria under Level 1 ($C_1 - C_4$), Level 2 ($C_5 - C_{13}$) and Level 3 ($C_{14} - C_{26}$) were evaluated qualitatively and quantitatively. The selection and the hierarchy of the criteria depended on **(1)** the availability of both quantitative and qualitative information, **(2)** data relating to the potential criteria, **(3)** relative importance in technology development and **(4)** available stakeholder information.

For the defined goal, four main criteria were selected under Level 1, namely:

C_1: **Economic**

The economic criterion C_1 considers the revenue that could potentially be generated (EUR/ kg treated waste) from discarded PCB. Estimating the potential revenue is straightforward, simply calculating the value of assayed metals in the waste material and their theoretical recovery efficiencies, and is inherently associated with uncertainties. An average metal concertation value is used for the revenue estimates, which translates into a uniform treatment alternative. In the weighing of this criterion, the process efficiencies were the main factors which differ considerably among the alternatives. Under this Level 1 criterion, costs (C_5), financial incentives (C_6) and revenues (C_7) were considered as significant Level 2 sub-criteria.

C_2: **Environmental**

The environmental profile is considered a significant Level 1 selection criterion, as the technology has potentially considerable impact on the environment. Many environmentally malign elements, such as corrosive acids, waste water, hazardous emissions, and waste heat are involved in the processing of waste materials for metal recovery. Moreover, mining activities of metal production from primary ores caused serious problems in the past which were overlooked or not foreseen before the initialization of the activities (Bakırdere et al., 2016). Similarly, there are raising concerns about the current state-of-the-art metal recovery practices in smelters (Akcil et al., 2015). Under C_2, the Level 2 sub-criteria operational environmental impact (C_8) and environmental costs (C_9) were considered.

C₃: Social

Waste management and resource recovery plays an important role in the social development of the society and the improvement of the quality of life (Guerrero et al., 2013). Social reputation, innovation potential, job creation and well-being improvement are considered as significant selection criteria. Social criteria are rather qualitative than quantitative, as the qualitative criteria measures against the criteria obtained from the public opinion. Social acceptability expresses the public opinion related to waste management and resource recovery from the stakeholder point of view. Under the Level 1 social criterion, social acceptance (C_{10}) and social benefits (C_{11}) are considered significant Level 2 sub-criteria.

C₄: Technical

This criterion essentially reflects the state-of-the-art of the applied technology and is here assessed based on the level of technology readiness level (TRL). Both alternatives considered in this chapter are technologically immature and far from full-scale application (Zhang and Xu, 2016). However, there are a significant number of research reports on both approaches (Ilyas and Lee, 2015; Jadhav and Hocheng, 2015). The level 2 sub-criteria, installation (C_{12}) and operation (C_{13}) are considered as relevant for this C_4 technical criterion. Several other MCA studies considered similar qualitative scales for technological maturity, i.e. operation and installation (Hallstedt et al., 2013; Kiddee et al., 2013; Ahmad et al., 2015).

4.2.4.3 *Establishing criteria weights*

The determination of the relative importance is inherently a cumbersome task, particularly, during early design stage when detailed information about the process is scarce. The selected criteria must be evaluated paying additional attention to potential conflicts (Herva and Roca, 2013; Ahmad et al., 2015). AHP techniques were used to determine the relative weightings of each criterion which was based on hierarchy, priority setting and logical consistency. Relative priorities were given to each element through pairwise comparisons 1–9, whereby 1 indicates equal, 3 moderate, 5 strong, 7 very strong and 9 extreme importance (Table 4-1). In pairwise comparison of the selected criteria, the analyzer shall determine if any given criterion (C_a) is of weak importance over the other (C_b), and accordingly assign a relative importance of for instance 2 to C_a. This translates into the value of the relative importance of C_b to C_a is ½. The quantified judgement of pairwise compassion is represented in a matrix, as explained below in 4.2.4.4.

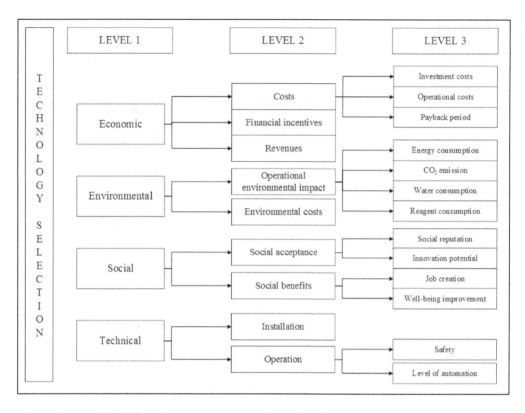

Figure 4-2: Analytical hierarchy process (AHP): the goal, Level 1, Level 2, and Level 3 criteria for metal recovery technology selection.

Table 4-2: Selected criteria expressed on a numerical scale and reciprocal value (Source: Saaty, 2001).

Scale of importance for comparison pair	Numeric rating	Reciprocal value
Extreme importance	9	0.111
Very strong to extreme importance	8	0.125
Very strong importance	7	0.143
Strong to very strong importance	6	0.167
Strong importance	5	0.200
Moderately to strong importance	4	0.250
Moderate importance	3	0.333
Equal to moderate importance	2	0.500
Equally important	1	1.000

4.2.4.4 *Construction of the evaluation matrix*

The matrix model defines a function from which the analytical representation of the goals extracted. The MCA goal can be expressed in matrix form as:

Criteria: $C_1, C_2, ..., C_n$

Weights: $w_1, w_2, ..., w_n$

Alternatives: A_1, A_2

$$A=\begin{bmatrix} A_1 \\ A_2 \\ ... \\ A_m \end{bmatrix} \begin{bmatrix} a_{11} & a_{12} & ... & a_{1n} \\ a_{21} & a_{22} & ... & a_{2n} \\ ... & ... & ... & ... \\ a_{m1} & a_{11} & ... & a_{mn} \end{bmatrix}$$

Where a_{mn} is the relative importance of criterion m to criterion n. Having computed the quantified pairwise comparisons on the pairs (C_m, C_n) as numerical entries w_{mn} in the matrix A, the set of numerical weights w_1, w_2, up to w_n are assigned. Finally, the alternatives' ranking is ordered and the best ranked alternative is calculated. The best alternative is weighted as per the matrix shown above, as the total weighted sum of all the selection criteria.

4.3 Results

4.3.1 Morphology of the discarded PCB

Binocular microscopic images and SEM micrographs revealed that the crushed PCB are heterogeneous, with varying sizes of particles, shapes and textures (Figure 4-1). Many particles showed rod-like or polygonal shapes, with flakes on the surfaces, typical for an end-product of a cutting mill (Yoo et al., 2009). The surfaces of the materials were angular in all board types, and the metallic particles were visually observable. Energy dispersive X-ray spectroscopy (EDS) analysis of the material showed that they contained substantial non-metallic silicon and carbon fragments, materials which are commonly used in PCB (Figure 4-1e). The pattern of the X-ray diffraction (XRD) analysis of the particulates of the material showed that the chemical composition showed no similarity to known natural minerals (data not shown).

4.3.2 Particle size of the boards

The discarded boards weighted (g/per PCB) 186.4 ±7.8, 64.2 (± 2.4), 178.4 (± 3.1), 40.8 (± 1.1) 144.4 (± 2.2) 297.5 (± 5.2), and 284.4 (± 4.4), respectively, for desktop computers, computer parts, laptops, mobile phones, computer boards without components, Telecom 1, and Telecom

2. The particle size distribution (PSD) from sieving test of the boards per PCB type is given in Table 3. After crushing between 3.24 – 6.56% of the material loss was measured (loss on comminution). This resulted in loss of material during crushing and transfer from the crusher to storage containers, and also during the sieving of the material.

The majority of the material mass was accumulated in the smallest particle size (<500 μm) in all PCB types with a variation between 51.7% to 71.8% of the total weight (Table 4-3). On the other hand, the weight ratio was the lowest for the coarsest particle size (1600 μm – 2500 μm) for all board types investigated, between 11.2% and 19.8%, and for telecom 2 boards and desktop computer boards without the components. The weight fraction of these two boards was very similar to the middle particle size fraction (500 μm – 1600 μm) with 83.2% and 92.4% weight of this particle size fraction. Ground waste material with the smallest particle size (<500 μm) was visually homogenous (Figure 4-1f), whereas coarser particle size showed differences (Figure 4-1h and Figure 4-1i). In the coarser particle sizes (1600 – 2500 μm and >2500 μm), particles of metals were easily detectable (Figure 4-1). Glass fibres are mainly enriched in the larger size fraction, while there are only tiny particles in the fine size fraction. However, visual interpretation did not reflect the analytical data, as the metal concentrations of the coarser particle sizes were in general higher than those of the smaller particles sizes as given below in 4.3.3.

4.3.3 Metal concentrations

The concentrations of Cu, Fe, Al, Ni, Zn, Pb, Au and of the boards per particle sizes (<500 μm, 500 – 1600 μm, and 1600 - 2500 μm) are given in Table 4-4. Copper (Cu) is the predominant material in all the discarded printed circuit boards (PCB) in varying concentrations. Maximum Cu content of 380.4 mg Cu/g PCB was in telecom 2 boards of particle size 1600 - 2500 μm, and the lowest with 93.4 mg Cu/g PCB in computer parts of particle size <500 μm. Following Cu, iron (Fe) and aluminium (Al) were the highest concentrated metals in most discarded boards except the two telecom boards. Desktop computer boards contained the highest iron (Fe) concentration compared to other boards. Telecom 2 contained the highest aluminium (Al) concentration in all particle sizes, however desktop computer boards also contained considerable amounts of this metal (Table 4-4). Gold (Au) was relatively less varied between particle sizes, however showed great variance between boards. Particularly telecom boards, and also laptop and mobile phone boards contained high concentrations of Au (244 – 320 ppm Au).

Other assayed metals, i.e. Cd, Co, Nd, La, Ce, Pt and Pd were below the detection limit (10 μg/g PCB).

4.3.4 Intrinsic value of the metals and total value of discarded printed circuit boards

The intrinsic values of the metals are given in Table 4-5. A breakdown of the individual metals and the total value of the discarded PCB were calculated. This value showed the potential revenue to be generated from the sustainable recovery of metals from discarded PCB. Cu and Au constituted the vast majority of the intrinsic value in all board types, varying between a combined total value of 92.6% to 98.6% (Table 4-5). The intrinsic values of Fe, Zn, Pb and Cr were insignificant, not surpassing 1% of the total value in any board type. Au is the most valuable metal and is the main driver of recycling of discarded PCB. Despite the dominance of Cu on a weight basis, it occupied only a small fraction of the total value of the laptop, mobile and telecom boards, but a considerable fraction of desktop computers and computer boards.

The total value of the PCB mainly depended on the Au concentration of in all the discarded PCB investigated. Computer boards had a smaller total value (2.36 EUR/kg PCB) due to a relatively low Au content of this type of boards. When the CPU units of desktop computers are removed prior to pretreatment, it influenced the total value of the board drastically. These units contain up to 1900 ppm gold (Birloaga et al., 2013). Thus, the content of Au was smaller to a 9.56 factor in the CPU-removed desktop computer boards compared to laptop boards. Telecom1 boards, on the other hand, included 10.2 times and 7.8% more Au than desktop computer boards and laptops, respectively. Also their Cu content was considerably higher than those of the other boards investigated. These boards are from high-end devices, and are operated at different conditions than regular desktop computer boards. Consequently, their total value was much higher (13.95 EUR/kg PCB) compared to desktop computer PCB (2.36 EUR/kg PCB).

4.3.5 Multi-criteria analysis and technology selection

Multi-criteria analysis gave the calculated weights of each alternative for technology selection, and the weights of each criterion and sub-criterion. Figure 4-3 shows the detailed results of AHP using the relative weights calculated for each Level 1 criterion, the weights for Level 2 sub-criteria and the selection for Level 3 sub-criteria. When all criteria are considered, biohydrometallurgy (A_2) had a higher marginal weight over hydrometallurgy (A_1) (0.535 > 0.465), despite the dominance of the latter on the most highly weighted Level 1 criterion: C_1

economic. The small marginal difference, however, showed the significance of the second best alternative, i.e. A_1 hydrometallurgy.

The Level 1 criteria had a weight of 0.466, 0.19, 0.172, and 0.172 for C_1 economic, C_2 environmental, C_3 social and C_4 technical, respectively (Figure 4-3). In the decision for technology development, economic factors are the most decisive and was rightfully reflected in the constructed AHP model. Furthermore, environmental and social criteria are also detrimental in environmental decision making, and they were equally weighted.

Among the Level 2 sub-criteria, biohydrometallurgy scored higher weight over hydrometallurgy in C_2 Environmental (0.716 > 0.284), C_3 Social (0.583 > 0.417), and C_4 Technical (0.583 > 0.417), which enabled biohydrometallurgy to have a higher overall weight over hydrometallurgy (Figure 4-3h, Figure 4-3i, Figure 4-3j). Under criterion C_1 economic, revenues received the highest weighing (0.54), while costs (0.297) and financial incentives received a relative lower weight (0.163). These results are in line with similar technology developments projects (Antonopoulos et al., 2014) at a an early stage of development. Under Level 1 criterion C_2 Environmental, operational environmental impact (C_9) had a much higher weight than environmental costs (C_{10}) as operational environmental impacts, e.g. energy consumption (C_{17}), CO_2 emissions (C_{18}), water consumption (C_{19}) and reagent consumption (C_{20}) have a higher projected cumulative impact rather than environmental costs, such as permits, and carbon emission costs. For criterion Level 1 C_3 Social, social acceptance and social benefits were equally weighted for technology selection.

Table 4-3: Screen size particle size distribution analysis of the discarded printed circuit boards from various devices.

PCB type	Desktops		Computer parts		Desktops without components		Laptops	
P.S (μm)	(g)	(%)	(g)	(%)	(g)	(%)	(g)	(%)
<500	62.2	64.9	51.7	54.1	64.2	67.9	64.1	68.4
500 - 1600	22.8	23.8	27.4	28.7	15.4	16.3	18.4	19.6
1600 - 2500	10.8	11.3	16.4	17.1	15.0	15.9	11.2	11.9
Total	95.8	100.00%	95.5	100.00%	94.6	100.00%	93.7	100.00%

PCB type	Mobile phones		Telecom 1		Telecom 2	
P.S (μm)	(g)	(%)	(g)	(%)	(g)	(%)
<500	59.8	56.0	61.8	71.8%	64.6	68.7%
500 - 1600	22.8	24.4	20.4	18.8%	17.6	20.8%
1600 - 2500	10.8	19.6	14.6	9.4%	14.6	10.5%
Total	93.44	100.00%	96.84	100.00%	96.76	100.00%

Table 4-4: Concentration of metals (mg/g PCB) in various parts of WEEE.

Metals	P.S (mm)	Desktops	Computer parts	Desktops w/o components	Laptops	Mobile phones	Telecom 1	Telecom 2
Cu	<500μm	176.7 ± 23.6	93.4 ± 7.2	163.6 ± 13.0	176.1 ± 18.6	230.1 ± 10.0	262.4 ± 16.6	305.4 ± 30.2
	500 - 1600 μm	241.6 ± 29.8	108.4 ± 4.6	222.4 ± 14.6	248.4 ± 21.4	274.4 ± 21.4	243.4 ± 44.4	312.6 ± 40.1
	1600 - 2500 μm	268.4 ± 24.2	112.2 ± 2.4	234.2 ± 20.1	266.5 ± 10.4	256.4 ± 18.9	293 ± 14.4	380.4 ± 37.4
Fe	<500μm	50.8 ± 5.4	20.4 ± 0.2	23.3 ± 2.6	37.8 ± 5.7	38.3 ± 3.1	11.4 ± 2.4	7.4 ± 0.2
	500 - 1600 μm	67.4 ± 9.2	26.4 ± 0.6	24.4 ± 1.8	44.4 ± 4.6	40.2 ± 4.1	26.6 ± 3.2	22.4 ± 2.2
	1600 - 2500 μm	75.8 ± 6.6	24.8 ± 2.1	26.6 ± 2.1	42.4 ± 3.8	46.6 ± 6.8	22.4 ± 2.6	22.2 ± 3.4
Al	<500μm	30.2 ± 1.6	15.4 ± 4.1	9.4 ± 6.6	19.8 ± 2.4	10.3 ± 4.3	49.0 ± 10.7	55.4 ± 4.2
	500 - 1600 μm	24.8 ± 3.5	12.2 ± 6.2	8.8 ± 1.8	24.8 ± 2.9	18.4 ± 1.4	26.6 ± 7.2	44.8 ± 4.5
	1600 - 2500 μm	30.4 ± 7.4	10.8 ± 4.1	5.6 ± 1.2	56.6 ± 10.1	26.6 ± 2.2	20.4 ± 2.3	41.9 ± 6.2
Ni	<500μm	5.0 ± 0.3	4.3 ± 0.7	2.6 ± 0.6	5.7 ± 0.7	11.5 ± 1.7	23.0 ± 4.8	12.2 ± 1.2
	500 - 1600 μm	15.0 ± 2.6	8.8 ± 1.1	0.6 ± 0.1	7.6 ± 1.2	8.4 ± 2.6	7.8 ± 2.4	23.2 ± 4.1
	1600 - 2500 μm	n.d.[7]	1.2 ± 0.4	n.d.	10.8 ± 1.9	10.6 ± 3.1	11.4 ± 1.4	24.8 ± 0.6
Zn	<500μm	5.3 ± 1.2	0.32 ± 0.02	2.9 ± 0.3	4.5 ± 0.7	3.0 ± 0.5	4.6 ± 0.6	4.2 ± 0.4

[7] n.d. : not detected

		Col 1	Col 2	Col 3	Col 4	Col 5	Col 6	Col 7
	500 - 1600 µm	6.4 ± 0.6	0.6 ± 0.01	4.1 ± 0.4	6.4 ± 0.6	4.0 ± 0.6	5.1 ± 0.5	5.4 ± 0.5
	1600 - 2500 µm	7.4 ± 1.2	1.4 ± 0.4	6.4 ± 0.2	5.8 ± 0.5	5.1 ± 0.9	5.5 ± 0.2	6.6 ± 0.6
Pb	<500µm	9.2 ± 1.2	4.1 ± 0.04	0.4 ± 0.01	2.2 ± 0.5	1.2 ± 0.5	4.2 ± 0.4	5.6 ± 0.5
	500 - 1600 µm	6.8 ± 0.6	3.6 ± 0.5	n.d.	2.4 ± 0.02	0.8 ± 0.01	2.1 ± 0.2	4.4 ± 0.6
	1600 - 2500 µm	8.1 ± 1.1	2.8 ± 0.02	0.8 ± 0.02	1.4 ± 0.01	n.d	1.2 ± 0.1	2.4 ± 0.2
Cr	<500µm	2.9 ± 0.1	1.4 ± 0.04	n.d.	1.0 ± 0.05	2.0 ± 0.2	40.6 ± 4	52.4 ± 11.1
	500 - 1600 µm	1.1 ± 0.4	2.1 ± 0.02	n.d.	1.2 ± 0.1	3.3 ± 0.3	54.4 ± 3.2	44.8 ± 12.4
	1600 - 2500 µm	2.1 ± 0.2	1.4 ± 0.01	n.d.	1.6 ± 0.08	5.4 ± 0.5	64.4 ± 1.8	66.4 ± 10.2
Au (ppm)	<500µm	31.4 ± 3.2	21.8 ± 0.5	30 ± 0.2	290.0 ± 26.0	297.4 ± 32.4	320.7 ± 17.1	298.2 ± 12.2
	500 - 1600 µm	44.2 ± 5.1	34.4 ± 2.8	32.4 ± 1.0	310.1 ± 21.0	310.4 ± 26.6	314.6 ± 28.2	300.4 ± 38.8
	1600 - 2500 µm	26.4 ± 4.7	28.6 ± 6.1	44.1 ± 4.4	244 ± 33.1	246.4 ± 41.4	286.1 ± 17.4	310.2 ± 24.2

n.d.: not detected

Table 4-5: Intrinsic value of copper and gold, and total value of the discarded boards.

Board type	Desktops	Computer parts	Desktops w/o components	Laptops	Mobile phones	Telecom 1	Telecom 2
Intrinsic value of Cu (%)	51.6	51.2	46.4	9.8	11.7	11.3	13.9
Intrinsic value Au (%)	41.3	41.3	51.6	89.0	86.9	85.8	83.2
Total intrinsic value of Cu and Au (%)	92.9	92.6	98.0	98.8	98.6	97.1	97.1
Total value of the board (EUR/kg)	2.36	1.18	2.4	12.30	12.74	13.95	13.66

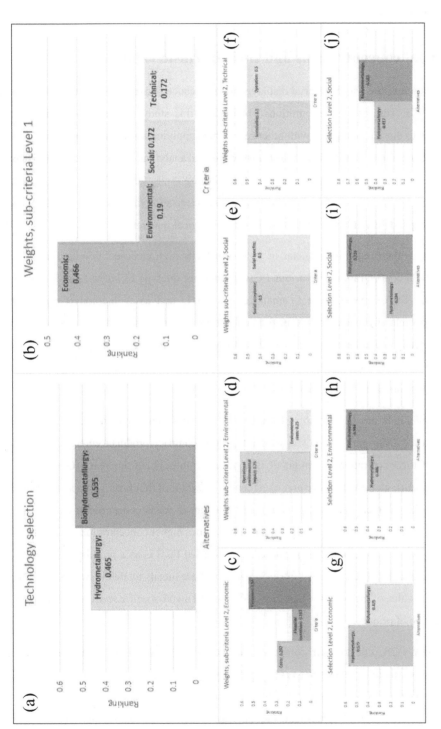

Figure 4-3: Multi Criteria Analysis results of (a) overall weighting of the two selected alternatives, (b) weights of the selected level 1 criteria (c, (d), (e), (f) weights of the sub criteria, C₁ economic, C₂ environmental, C₃ social, and C₄ technical respectively; and (g), (h), (i), (j) the weights of the alternatives for the level two sub-criteria.

4.4 Discussion

4.4.1 Metal concentrations and their intrinsic value

Characterization of the waste material displayed an abundance and a large variety of valuable metals in PCB. The metal concentrations obtained in this study (Table 4-4) are in good agreement with those investigated with other researchers applying a similar methodology with regard to pre-treatment, particle size, digestion method, and analytical method (Veit et al., 2006; Chancerel and Rotter, 2009; Kasper et al., 2011; Gurung et al., 2013). Variations in metal composition of PCB may be attributed to different characterization methods applied, changing manufacturing and design techniques as well as technological innovations.

Copper (Cu) is the predominant metal in PCB due to its high electrical conductivity. The variation of Cu concentrations in various PCB stems from the type of board (single layer or multi-layer), year of manufacture (Yamane et al., 2011) and assay method (Chapter 3). Moreover, particle size played a role in the concentration of metals, as the concentration of individual metals varied between particle sizes of the same PCB type (Table 4-4). Also, more advanced boards, i.e. the ones manufactured at an earlier date, included higher metal concentrations compared to the older ones.

Boards of all kinds have observably decreased in size and multi-layered boards are more commonly used (Marques, 2013). As a results, more compact products emerged, thus an increase in Cu concentration in similar PCB is observed throughout the years (Ghosh et al., 2015). Gold (Au), along with other precious metals, is used as a thin layer contact material in PCB due to its chemical stability. A significant amount of Au is concentrated on the CPU (Birloaga et al., 2013). Discarded PCB contains many times higher concentration of Cu and Au than natural ores (Akcil et al., 2015). Therefore, discarded PCB is as a promising secondary source of Cu and Au. The combined intrinsic value of these metals totaled up to 96.8% of the total potential value of the boards. Therefore, efficient and metal-specific recovery of these two metals must be of priority.

Mobile phone PCB, due to their compact size and design, have higher concentrations of relatively more valuable metals such as Cu, Ni, Cr and Au, but lower concentrations of Fe and Pb. This is a result of advancing board technologies and soldering techniques, as well as restrictions on the usage of certain substances, e.g. Pb as per the restrictions dictated by a number of directives, such as the Restriction of Hazardous Substances Directive (RoHS Directive 2002/95/EC) by the European Commission.

For the metal assay of PCB, microwave-assisted digestion is a rapid and accurate technique. Measurement of metals in environmental samples using microwave-assisted digestion methods that yield virtually-total metal content while avoiding the use of hydrofluoric acid (HF) on a siliceous dominated matrix (Hassan et al., 2007). Recoveries of other elements such as Cr and Ni compared well with 'near-total' recoveries yielded by conventional (non-MW assisted) acid digestion methods (Chapter 3).

4.4.2 Importance of particle size in the metal extraction step

The particle size has a negative correlation with metal removal efficiency, and high particle sizes may lead to a poor metal extraction efficiency (Young and Veasey, 2000). This applies to both hydrometallurgical or biohydrometallurgical processing of WEEE, in which a physical contact with the leachant on the material is required. As the surface area is increases by decreasing particle size, the contact between the metal-bearing waste particles and the leachant in the leaching medium increases, thus leading to a higher metal extraction efficiency. Moreover, microbial cells attach to the waste material, which enhances the metal removal efficiency and bioleaching rate in biohydrometallurgical processes (Silva et al., 2015). However, over-crushing and excessive sieving cause either loss on comminution (LoC) or accumulation of the material on the sieve (Chancerel et al., 2009; Yoo et al., 2009). Current design for crushers lead to a loss of metals during pretreatment (Ueberschaar and Rotter, 2015). Thus, a balance between an appropriate amount of crushing and selection of the correct particle size is required. A few studies investigated the advanced pretreatment of the waste PCB material using electrostatic (Eddy-current) and gravity based techniques (Ruan et al., 2014). These techniques might prove useful provided that minimal loss of valuable metals is achieved. Nonetheless, manual crushing and sieving is not realistic in a full-scale plant, and a novel strategy for the optimization of the physical pretreatment step is required.

4.4.3 Technology selection for metal recovery

Evaluation of multiple economic, environmental, social and technological criteria, and the consequences of full-scale application enable to take a strategic decision in technology selection at an early stage of development (Cundy et al., 2013). Multi-criteria analyses (MCA) are increasingly applied to emerging waste management and resource recovery technologies with the aim to provide an analytical decision-support tool (Huang et al., 2011). Analytical hierarchical process (AHP) method is useful when a multi-dimensional and complex technological selection goal is concerned, which typically involves a range of conflicting criteria featuring different forms of data and information. Such an assessment at an early stage of development allows for investigation and integration of the interests and objectives of multiple criteria through the input of both quantitative and qualitative data.

Application of MCA provides support in technology selection with various factors. Such decision making support methods may be a vital component in substantiating the metals recovery scenarios and to facilitate the process of selection of different technologies. The output of the AHP method for technology selection at an early stage of research deals with the complexity of the multi-factor and criteria setting by providing throughput communicable information. It allows objectivity and inclusiveness of various dimensions of sustainable development. Multi-criteria decision-making models, like the one developed here, can be used to assess, compare and rank different metal recovery technologies in a transparent way based on a comprehensive set of technical, environmental and social criteria. Uncertainty is inherent in most MCAs and is not easily remediated. The uncertainty in the input information going into a MCA can be reduced through careful evaluation of the available data sources to ensure that e.g. the inputs are representative of the study context and that the underlying assumptions behind the inputs are consistent.

In this study, biohydrometallurgy is conclusively found to perform well when the selected 4 Level 1 criteria, and the 22 Level 2 and Level 3 sub-criteria are concerned. This technology is an emerging and promising alternative with low impacts on amenity, high public acceptance, high potential for innovation and still have the potential to deliver adequate technical performance. Despite its relative frail economic strength, which may explain the relatively slow uptake of these technologies, biohydrometallurgy alternative should be offered great incentives owing to its eco-innovation potential, environmental friendliness and vast improvement potential. Although conventional hydrometallurgical routes could perform economically better, owing to their established sound processes and higher efficiencies, biohydrometallurgical routes are generally better suited for the Level 1 criteria considered other than economic (C_1) factors Figure 4-3. Thus, a shift towards an increasing use of biohydrometallurgical processes for metal recovery can thus be expected. This does, however, not imply the negligence or takeover of the established conventional chemical technologies.

Nevertheless, the selection of biohydrometallurgy as the best alternative must be judged cautiously. This is a novel technology for which a small number of industrial applications for primary ores exist up to date, and many fundamental concepts remain unexplored. Besides, it was not initially conceived as a process for the treatment of secondary waste, but for the primary metal-bearing minerals, given the natural ability of the microbes to process metal ores. Further, it must be remarked that only social and environmental criteria were the main influence to obtain this ranking. Nevertheless, the selection of one or another metal recovery alternative would be conditioned by emerging new technologies, the resources availability as well as by social, political or economic aspects not being considered here. Nevertheless, the selection of one or another metal recovery alternative would be conditioned by the technological readiness level (TRL) and the resources availability as well as by social, political or economic aspects not being considered here.

4.5 Conclusions

A modified metals characterization method for discarded PCB was applied to various waste material from various sources. The discarded PCB contained 9.3% - 38.4% (by weight) of Cu, and 21 - 320 ppm of Au, along with considerable concentrations of Fe, Al, Ni, Zn, and Cr. Cu and Au constituted a total of 91.3 and 96.8% of the total value of the waste material and the discarded PCB is a very promising secondary source of these two metals. The concentration of Au was the decisive factor in the value total of the discarded boards and its concentration was the main factor in the total value of the boards. The concentrations of the other metals were considerably high as well, however their fraction was economically neglectable, not exceeding 1% of the total value. On the other hand, their presence and abundance is of importance in the development of a sustainable metal-selective recovery technology.

Extraction of metals form the waste material is an essential process in metal recovery from WEEE. This MCA study revealed that the biohydrometallurgical route is slightly preferred over hydrometallurgical routes (0.535 > 0.465) in technology selection for sustainable metals recovery from WEEE. Despite its relative poor performance in the most relevant economic criterion, its overall weight was slightly higher, owing to its better environmental and social profile. Conclusively, the application of MCA in technology selection is a useful method. It is expected that the use of this technique and complexity of the analysis will increase in future.

Chapter 5.

Bioleaching of copper and gold from discarded printed circuit boards

This chapter is based on: Işıldar, et al. Two-step bioleaching of copper and gold from discarded printed circuit boards (PCB). Waste Management (2015), http://dx.doi.org/10.1016/j.wasman.2015.11.033

Abstract

An effective strategy for environmentally sound biological recovery of copper and gold from discarded printed circuit boards (PCB) in a two-step bioleaching process was experimented. In the first step, chemolithotrophic acidophilic *Acidithiobacillus ferrivorans* and *Acidithiobacillus thiooxidans* were used. In the second step, cyanide-producing heterotrophic *Pseudomonas fluorescens* and *Pseudomonas putida* were used. Results showed that at a 1% pulp density (10 g/L PCB concentration), 98.4% of the copper was bioleached by a mixture of *A. ferrivorans* and *A. thiooxidans* at pH 1.0-1.6 and ambient temperature ($23 \pm 2°C$) in 7 days. A pure culture of *P. putida* (strain WCS361) produced 21.5 (± 1.5) mg/L cyanide with 10 g/L glycine as the substrate. This gold complexing agent was used in the subsequent bioleaching step using the Cu-leached (by *A. ferrivorans and A. thiooxidans*) PCB material, 44.0% of the gold was mobilized in alkaline conditions at pH 7.3-8.6, and 30°C in 2 days. This study provided a proof-of-concept of a two-step approach in metal bioleaching from PCB, bacterially produced lixiviants.

5.1 Introduction

Waste electrical and electronic equipment (WEEE) is generated at an exponentially increasing rate. Global WEEE generation was estimated to reach 41.8 million tons in 2014, and forecasted to rise to 50 million tons in 2018 (Baldé et al., 2015). In the EU-27, 19.1 kg per inhabitant is generated, a total of 9.8 million tons in 2013 (StEP, 2015). It makes up 2–3% (up to 8% in the developed economies) of municipal waste (Widmer and Oswald-Krapf, 2005; Robinson, 2009).

Various substances found in WEEE, i.e. heavy metals (copper, chromium, lead, nickel, zinc, etc.), brominated flame retardants (BFRs), polybrominated diphenyl ethers (PBDEs), and chlorofluorocarbons (CFCs) pose a serious threat to the environment and public health when improperly disposed (Tsydenova and Bengtsson, 2011). In the EU, only 33% of WEEE is reported to be treated properly (Torretta et al., 2013). In addition, 13% of the collected WEEE is not reported as being processed, presumably it has been treated under substandard conditions (Breivik et al., 2014).

In addition to being very hazardous, WEEE, and in particular printed circuit boards (PCB), contain high concentrations of metals (Ghosh et al., 2015). Previous studies have reported concentrations in the range of (g/kg): 100 - 350 Cu, 10 - 100 Fe, 10 - 50 Al, 1 - 10 Ni, along with 0.01-0.035 Au, 0.02-0.150 Ag and 0.01-0.12 Pd (Veit et al., 2006; Cui and Forssberg, 2007; Yamane et al., 2011). Metal recovery from PCB, a secondary source, can support the conservation of primary sources and prevent environmental degradation while contributing to a transition to a circular economy (Wang and Gaustad, 2012).

Various metal recovery methods from WEEE are available, including smelting, acid leaching and bioprocessing, each of these methods having their own limitations (Cui and Zhang, 2008; Ilyas and Lee, 2014a). Recovery of metals is not easy due to the prevalence of many different types of substances integrated into the PCB (Tanskanen, 2013). Bioprocessing, including a number of microbially induced processes e.g. bioleaching, bioreduction and biosorption, attract interest for metal recovery from waste materials due to their environmental-friendly and cost-effective nature. A wide range of chemolithotrophic (Xiang et al., 2010), heterotrophic (Chi et al., 2011), and thermophilic (Ilyas et al., 2007) bacteria as well as fungi (Brandl et al., 2001) have been tested for their ability to mobilize base metals, e.g. Cu, Zn, Fe, Ni, and precious metals, e.g. Au, Ag, Pd, Pt from discarded PCB through various mobilization mechanisms. Bioleaching media are inoculated with cultures of chemolithotrophic acidophilic iron- and sulfur-oxidizing microorganisms where biogenic sulfuric acid and ferric iron act as lixiviants via respectively, acidolysis and redoxolysis bioleaching mechanisms (Lee and Pandey, 2012). In these mechanisms, metals are mobilized to their ionic state in aqueous solution by proton attack via formation of acids (acidolysis) or oxidation/reduction reactions (redoxolysis). Heterotrophic cyanide producing microorganisms produce free cyanide (CN^-) which complexes and mobilizes metals through a disparate dissolution mechanism, termed complexolysis (Brandl, 2008).

In this study, bioleaching of Cu and gold from discarded PCB in a two-step process was studied, in which redoxolysis, acidolysis and complexolysis mechanisms are involved. A number of studies on single-step direct copper (Zhu et al., 2011; Bas et al., 2013) and gold (Chi et al., 2011; Natarajan and Ting, 2014) bioleaching have been reported. In order to further improve the yield of metal mobilization, as well as decrease lixiviant consumption, a novel two-step approach is proposed based on the different chemical properties and leaching mechanisms of base and precious metals. It builds on earlier studies, that showed that the removal of copper by biooxidation was found to improve gold recovery (Ting and Pham, 2009). Specific objectives of this work are **(a)** to establish an effective metal recovery strategy for discarded PCB, **(b)** to determine the efficiency of the successive application of acidophilic and cyanogenic bacteria to mobilize metals from these waste materials, and **(c)** to study the optimal process parameters for this two-step bioleaching of metals.

5.2 Materials and methods

5.2.1 Source and preparation of the waste material

Discarded electronic devices, namely desktop computers, laptop computers and cell phones, were collected from the IT department of UNESCO-IHE (Delft, the Netherlands) and SIMS recycling (Eindhoven, the Netherlands). Batteries, heat sinks and microprocessors were removed from the PCB of desktop computers and laptops. Parts of desktop computers, i.e. video cards and sound cards, were manually removed in order to determine their metal concentrations. Components of the boards such as connectors, capacitors, and integrated chips were removed using a thermal gun (Gamma, HG2000E, the Netherlands). The scrap PCB of various devices were grouped into 5 categories; namely desktop computer boards, computer parts, desktop computer boards without components, laptop computer boards, mobile phone boards.

The PCB were washed with distilled water, cut manually, crushed by a cutting mill (Retsch 2000, Germany) and sieved (Fritsch, UK) to a particle size smaller than 500 μm. Ground PCB material was washed with distilled water and sterilized by tyndallization, heating the material three days in succession to 95°C and holding it at that temperature for 15 minutes, prior to each step of the bioleaching experiments.

5.2.2 Microorganisms and cultivation

Acidophilic strains of *Acidithiobacillus ferrivorans* (formerly known as *Acidithiobacillus ferrooxidans,* DSM 17398) and *Acidithiobacillus thiooxidans,* (DSM 9463) were ordered from the Leibniz Institute (DSMZ), Braunschweig (Germany). Cyanide producing strains *Pseudomonas putida* (WSC361) and *Pseudomonas fluorescens* (E11.3) were kindly provided by Dr. Peter Bakker from Utrecht University (the Netherlands).

The *Acidithiobacillus* strains were grown in a mineral medium containing (g/L): $(NH_4)_2SO_4$ (2.0), $MgSO_4 \cdot 7H_2O$ (0.25), KH_2PO_4 (0.1), KCl (0.1), $FeSO_4 \cdot 7H_2O$ (8.0), and S^0 (10.0) (Brandl et al., 2001). The pH was set to 2.5 with sulfuric acid. The cultures were inoculated with 5% (*v/v*) in 100 mL growth medium in 300 mL Erlenmeyer flasks and incubated at 30°C at a rotation speed of 150 rpm for 7 days prior to the bioleaching experiments.

Cyanogenic strains were grown in nutrient broth containing (g/L): meat extract (1.0), yeast extract (2.0), peptone (2.0), and NaCl (5.0) for subculturing and in glycine-supplemented medium (Shin et al., 2013) for cyanogenic activity and bioleaching tests. The cultures were subcultured with 1% (*v/v*) in 100 mL growth medium in 300 mL Erlenmeyer flasks and incubated at 30°C at a rotation speed of 150 rpm.

Grown cells were enumerated using the viable count method, with a setup consisting of 9 consecutive tubes and up to 10 dilutions followed by the spread plate method (Starosvetsky et al., 2013). Acidophiles were plated on *Thiobacillus* agar (g/L): $(NH_4)_2SO_4$ (0.4), $MgSO_4 \cdot 7H_2O$ (0.5), $CaCl_2$ (0.25), KH_2PO_4 (4.0), $FeSO_4 \cdot 7H_2O$ (0.01), $Na_2S_2O_3$ (5.0) and agar (12.5). The activity of *Acidithiobacillus* strains was monitored indirectly by measuring pH in the bioleaching medium. Cyanogenic bacteria were plated on peptone agar (g/L): peptone (10.0), NaCl (5.0) and agar (12.0) as described by Campbell et al. (2001). Growth of the *Pseudomonas* strains was monitored by measuring optical density (OD) at 600 nm.

5.2.3 Copper bioleaching experiments

Various concentrations (0.5, 1, 2.5, 5%, *w/v*) of ground PCB waste (<500 μm) were added to 100 mL of bioleaching medium containing active growing cultures ($1.2 \pm 0.4 \times 10^8$ CFU/mL) in 300 mL Erlenmeyer flasks on an orbital shaker (Brunswick Innova 2000, USA). Operating conditions of temperature, agitation rate, and leaching period were set to ambient temperature ($23 \pm 2°C$), 150 rpm, and 480 h, respectively. Control experiments were carried out with non-inoculated bioleaching medium. pH, oxidation reduction potential (ORP) and cu concentration of the leachate solution were monitored periodically. Each flask experiment was done in triplicate.

5.2.4 Gold mobilization experiments

The residue of the first bioleaching step was collected, filtered (cellulose filter, Whatmann, UK) left to dry overnight, and sterilized by tyndallization as described in 5.2.1. Active growing cultures ($2.1 \pm 0.5 \times 10^9$ CFU/mL) of *Pseudomonas putida* and *Pseudomonas fluorescens* were inoculated with 1% (*v/v*) growth medium described in 5.2.2 with supplementation of various concentrations of glycine (5, 7.5, 10 g/L). The sterilized material was then subjected to secondary bioleaching by adding the dried residue of the first bioleaching step to cultures of cyanogenic bacteria at the point of maximal cyanide generation. Operating conditions of temperature, agitation rate, and leaching period were set to 30°C, 150 rpm, and 120 h, respectively.

Chemical cyanide leaching experiments from virgin material and residues were carried out by adding the material at various pulp densities (0.5, 1, 2.5%, w/v) and cyanide concentrations (10, 25, 50, 100, 1000 mg/L). Operating conditions of temperature, agitation rate, and leaching period were set to ambient temperature (23 ± 2°C), 150 rpm, and 60 h, respectively. Each flask experiment was done in duplicate.

5.2.5 Characterization of discarded PCB

A modified EPA method (3052-1996 Microwave assisted acid digestion of siliceous and organically based matrices) was used for characterization of the PCB material. 0.5 g of the crushed sample was added to 12 mL of nitro-hydrochloric acid, a mixture of concentrated hydrochloric acid and nitric acid, at a ratio of 3:1 (v/v), respectively, in a laboratory microwave accelerated reaction unit (CEM Duo Temp Mars 5, USA). The mixture containing acid and the sample was gradually heated up to 175°C in 5.5 min and remained at 175°C for 9.5 min in fluorocarbon polymer microwave reaction vessels (CEM, MarsXpress, USA). After cooling, the vessel contents were filtered (Whatmann, G/C) twice, serially diluted, and analyzed for metal concentrations. The final solid residues were also examined using optical microscopy (BX63, Olympus, Japan), binocular microscopy (SZ61, Olympus, Japan) as well as scanning electron microscopy (JSM-6010LA, JEOL, Japan).

5.2.6 Detection of cyanide production by the cultures

Detection of the cyanide production ability of *Pseudomonas* cultures was done using a modified colorimetric method (Knowles, 1976). Nutrient agar supplemented with glycine was poured into petri dishes and inoculated with active growing *Pseudomonas* cultures. Sterile filter paper soaked in 0.5% picric acid solution was fixed to the underside of the Petri dish lid. The dishes were sealed with paraffin film and incubated at 30°C. Biologically produced free cyanide ions (CN^-) were measured by a potentiometric method with an ion selective electrode (ELIT 8291, Nico 2000, UK) and titration against standard $AgNO_3$ solution as described by Zlosnik and Williams, (2004). Measurements of CN^- were made by connecting the electrode to a pH meter (WTW 341i, Germany) set to read on mV. Calibrations (0.1-1000 mg/L) were carried out prior to CN^- measurements. To measure biogenic cyanide, cultures were grown into stationary phase and aliquots of 5 mL were sampled from the medium. Supernatant was collected by centrifugation at 12,000 g for 10 min at 4°C. To minimize contamination of the electrode, the supernatant was passed through a membrane filter with pore size 0.45 µm prior to CN⁻ measurement. The pH of the supernatant was set to around pH 12 using 5 M NaOH (10 mL/L) to bring it within the optimal pH range of the electrode (pH 11–13).

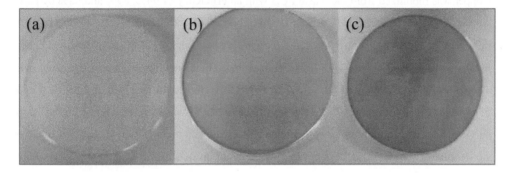

Figure 5-1: *Pseudomonas putida* strains spread-plated on nutrient medium supplemented with glycine in pet dishes with lids including filter papers soaked in 0.5% Picric acid after(a) 24 hours (b) 48 hours and (c) 72 hours.

5.2.7 Analytical methods

Metal analyses were performed by inductively coupled plasma mass spectrophotometry (ICP-MS) (X series 2, Thermo Scientific, USA) for waste characterization as described by Kolias et al. (2014). Periodic cu measurements were carried out by atomic absorption spectrophotometer (AAS) (Varian 200, USA) at 324 nm wavelength. Prior to analysis, the samples were centrifuged at 12,000 g for 10 min, acidified and stored at 4°C. Gold measurements of biological samples were carried out by sampling aliquots of 5 mL from the bioleaching medium, centrifugation at 12,000 g for 10 min at 4°C and digestion with nitro-hydrochloric acid at 95°C for one hour. The evaporation loss was compensated by adding ultrapure water to the aliquot. The final digestate was acidified by 1% HCl (37%) and diluted prior to Au measurements.

pH and ORP was measured by Ag/AgCl reference electrodes (SenTix 21, WTW, Germany and QR481X, Qis, the Netherlands) as described by Plumb et al. (2008) and Zhao et al. (2015) respectively.

5.3 Results

5.3.1 Copper bioleaching by acidophiles

Copper was mobilized from the waste PCB with an efficiency of 94%, 89% and 98% by pure cultures of *A. ferrivorans*, *A. thiooxidans* and a mixture of both, respectively, at 1% (10 g/L) pulp density. Figure 5-2 shows the bioleaching profile of cu and pH at 1% pulp density. Higher loads of waste material showed a lower bioleaching efficiency coupled to an increase of pH. The bioleaching efficiency of cu from waste PCB at various pulp densities of 0.5, 1, 2.5 and 5% (*w/v*) is given inFigure 5-3. A pulp density below 2.5% was more efficient with regard to cu yield in solution and mobilization efficiency. Very low cu mobilization was observed in the sterile control which could be attributed to the addition of sulfuric acid in the bioleaching medium (data not shown).

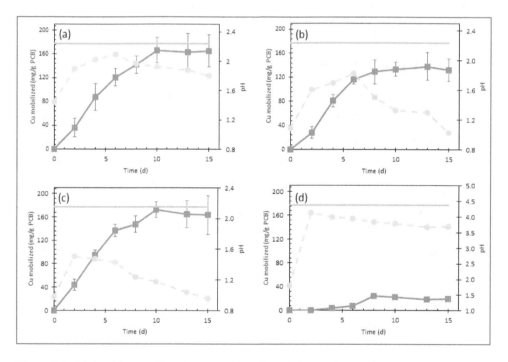

Figure 5-2: Bioleaching profile of copper (■) and pH (●) by cultures of (a) *A. ferrivorans*, (b) *A. thiooxidans*, (c) mixture of *A. ferrivorans* and *A. thiooxidans* and (d) non-inoculated control at 1% pulp density. Standard errors are within 0.2 pH units. Straight lines indicate 100% copper mobilization.

PCB are inhibitory on bacterial activity due to its various hazardous compounds such as heavy metals, phenols, and BFRs (Liang et al., 2010). In order to mitigate PCB toxicity on the bacteria, as well as to favor bioleaching conditions, a pre-growth method was applied. Acidophiles were first incubated in the bioleaching medium in the absence of waste material until the cultures were established for optimal bioleaching conditions of low pH and high oxidation reduction potential (ORP). Subsequently, the PCB was added in various concentrations.

In bioleaching systems, bacterial activity can be monitored by measuring pH and ORP. These two parameters indicate the involvement of acidolysis and redoxolysis mechanisms. The pH decreased with the oxidation of elemental sulfur and formation of sulfuric acid indicating acidophilic activity (Figure 5-2). At the end of the pre-growth period of the pure cultures *A. ferrivorans*, *A. thiooxidans* and the mixed co-culture, the pH had dropped to 1.45 ± 0.1, 1.1 ± 0.1 and 0.95 ± 0.1, respectively (Figure 5-2a, b and c).

The pH profile at various pulp densities (0, 0.5, 1, 2.5 and 5%) is shown in Figure 5-3b. The pH observably increased after addition of the waste material due to its alkaline nature (Brandl et al., 2001). At pulp densities of 2.5% and higher, the pH did not drop down to acidic levels (pH<2.5) in which acidophiles thrive optimally. Microorganisms thrived well at a pulp density of 1% and below. The ORP increased with formation of sulfuric acid and oxidation of ferrous to ferric iron indicating acidophilic activity. The ORP profile showed a similar trend to pH, where a response was observable with the addition of waste material followed by a steady state condition where bacteria retained optimal bioleaching conditions (Figure 5-3c).

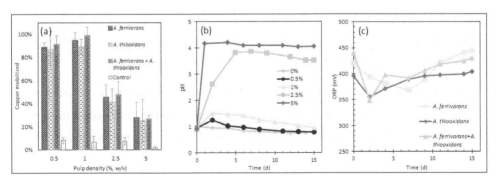

Figure 5-3: Final copper mobilization efficiency of the bioleaching cultures at various pulp densities (a), pH (b) and ORP (c) profile at 1% pulp density. Standard errors are within 0.2 pH units and 25 mV for (b) and (c) respectively.

5.3.2 Morphology and chemical structure of the particulates

Binocular microscopic images and SEM micrographs of the crushed PCB revealed that the material is heterogeneous, with particles of varying sizes, shapes and textures. Many particles showed rod-like or polygonal shapes, with flakes on the surfaces, typical for an end-product of a cutting mill (Yoo et al., 2009). After bioleaching, the surfaces of the material had eroded, and removal of metallic particles was visually observable. Energy dispersive X-ray spectroscopy (EDS) analysis of the virgin material showed that they contained substantial non-metallic silicon and carbon fragments, materials which are used commonly in PCB.

The pattern of the X-ray diffraction (XRD) analysis of the particulates of the virgin material, and after the first bioleaching step did not match any known natural mineral from the database (data not shown). Mass balances of the metals were performed by assaying the concentrations in residues after the first and second bioleaching steps as per the method explained in section 5.2.5.

5.3.3 Cyanide production by *Pseudomonas* strains

The cyanogenic ability of the strains was confirmed by the colorimetric picric acid test with development of orange to red color from yellow (data not shown). In these tests *P. putida* (WCS 361) showed slightly more intense color than *P. fluorescens* (E11.3). This was confirmed by relatively higher cyanide production by *P. putida* than by *P. fluorescens* at 21.5 ($\pm$ 1.4) and 15.5 ($\pm$ 2.4) mg/L, respectively, under optimal precursor concentrations, as shown inFigure 5-4a and Figure 5-4b. The biogenic cyanide yield by *P. putida* was correlated with the glycine concentration, whereas *P. fluorescens* provided a lower yield at a high glycine concentration, indicating an inhibitory effect of glycine above 7.5 g/L for this strain.

Cyanide production reached a maximal level at the late exponential and early stationary phases (Figure 5-4a and b). Subsequently, the cyanide concentration dropped at the late stationary and decay phases. The cyanide concentration was below the detection limit in controls without glycine addition.

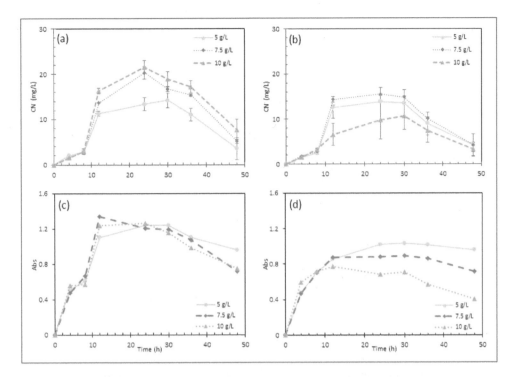

Figure 5-4: Cyanide production (a and b) and growth (c and d) of *P. putida* and *P. fluorescens*, respectively, at various glycine (5, 7.5, 10 g/L) concentrations. Standard errors for absorbance are within 0.15 units.

5.3.4 Mobilization of gold

A similar waste material addition strategy as for the first step was followed where the Cu-leached residue was added at the time of maximal cyanide production by the bacteria. Gold leaching using biogenic cyanide and chemical cyanide at various pulp densities is shown in Figure 5-5. At 0.5% pulp density, *P. putida* strains achieved the highest gold recovery of 44%, as shown in Figure 5-5d. Gold was not detected in the sterile control throughout the entire bioleaching period, indicating that biogenic cyanide was the sole gold mobilization agent.

In parallel, gold leaching experiments with chemical cyanide showed a direct positive correlation of the cyanide concentration with the gold mobilization efficiency (Figure 5-5c). A cyanide concentration at 100 mg/L (3.84 mM) leached all the gold (113.1 ± 11.2%) from the residual waste material. Direct cyanidation of the non-Cu-leached material with the same cyanide concentration (100 mg/L) at 1% pulp density leached 27.2 (±5.9)% and 36.0 (±7.2)% of the Au and Cu, respectively (data not shown).

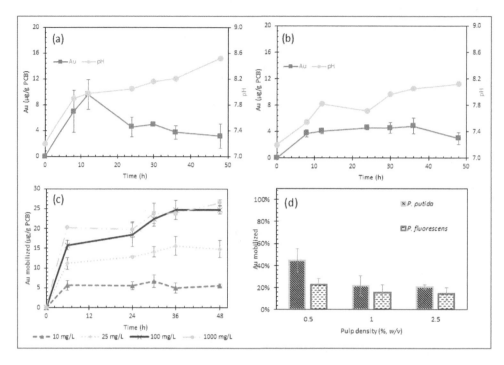

Figure 5-5: Gold mobilization of P. putida (a), P. fluorescens (b) at 0.5% pulp density, chemical cyanide (c) at 1% pulp density and efficiency. Standard errors are within 0.25 pH points for (a) and (b).

5.4 Discussion

5.4.1 Copper bioleaching from PCB by acidophiles

In bioleaching tests, a pregrowth strategy was applied, where microorganisms are cultivated in the absence of the waste material, allowing the cultures to establish and maintain optimal bioleaching conditions. This showed a shortened bioleaching duration and no lag phase before cu dissolution as opposed to studies with direct inoculation (Xiang et al., 2010; Zhu et al., 2011). Cu bioleaching showed a logarithmic increase trend followed by a decrease in the concentration curve, typical for bioleaching systems (Wang et al., 2009; Liang et al., 2013).

Bioleaching of metals from non-sulfide wastes involves direct and indirect leaching mechanisms by biogenic ferric iron (Fe^{3+}) and sulfuric acid (H_2SO_4). Attached cells play a major role in bioleaching of cu from PCB (Nie et al., 2015a; Silva et al., 2015). The role of the acidophiles in these processes is to catalyze the oxidation of ferrous (Fe^{2+}) to ferric (Fe^{3+}) iron and elemental sulfur (S^0) to sulfuric acid:

$$4Fe^{2+} + O_2 + 2H^+ \rightarrow 4Fe^{3+} + 2OH^-$$ (5-1)

$$S^0 + 1.5O_2 + H_2O \rightarrow 2H^+ + SO_4^{2-}$$ (5-2)

The biogenic ferric iron (redoxolysis) and proton acidity (acidolysis) leaches the cu from the PCB under low pH and high ORP conditions:

$$Cu^0 + 2Fe^{3+} \rightarrow Cu^{2+} + 2Fe^{2+}$$ (5-3)

$$Cu^0 + H_2SO_4 + 0.5O_2 \rightarrow Cu^{2+} + SO_4^{2-} + H_2O$$ (5-4)

The alkalinity of the PCB and the oxidation of Fe^{2+} into Fe^{3+} resulted in an increase of pH and decrease of ORP. Evidently, bioleaching of PCB via redoxolysis and acidolysis mechanisms is an acid consuming process. Moreover, biologically induced conversion of Fe^{2+} into Fe^{3+} indicates a cyclic process as given in equations (1) and (3).

A co-culture of iron and sulfur oxidizing acidophiles showed more extended cu mobilization than the pure cultures for all the selected pulp densities (Figure 5-3a). This suggests a twofold advantage of involvement of both leaching mechanisms, redoxolysis and acidolysis, as well as a positive cooperative bioleaching mechanism when the acidophiles coexist in the bioleaching co-culture. An increased bioleaching performance of acidophiles in communities has been shown for bioleaching of minerals (Spolaore et al., 2011) and waste PCB material (Liang et al., 2010).

Pulp densities of 2.5% or higher severely affected the viability of the microorganisms. Likely, the inhibition of microbial activity by an unidentified component of the PCB compromised the rate of metal mobilization. It has been argued that the metal ions leached from the waste material play the largest role in microbial inhibition by PCB components (Ilyas et al., 2007). On the other hand, the inhibitory effect may also originate from the organic fraction of the PCB (Zhou et al., 2013). Toxic effects of organic compounds on acidophiles, including aromatics and brominated flame retardants, lead to decreased iron oxidation rates, and suppression of the CO_2 fixation enzymes (Tabita and Lundgren, 1971; Stapleton et al., 1998). Moreover, *Acidithiobacillaceae* can sustain concentrations up to 800 mM (20.8 g/L) of cuprous ions (Orell et al., 2010), the most abundant cation in the PCB leachate solution investigated. This highlights the likely toxicity of organics on the acidophiles.

5.4.2 Cyanide production and gold bioleaching by *Pseudomonas* strains

Our results are in good agreement with earlier literature reports on the cyanide concentration and the timeframe of the cyanide production (Campbell et al., 2001; Pradhan and Kumar, 2012; Ruan et al., 2014). Cyanide is produced as a secondary metabolite during the oxidative decarboxylation of glycine (Blumer and Haas, 2000). Biogenic cyanide production by *P. putida* is positively correlated with the glycine concentration, whereas *P. fluorescens* provided a lower yield at high glycine concentration, as shown in Figure 5-4b, indicating the inhibitory effect of glycine at 10 g/L for this strain. In addition, absorbance measurements showed a lower cell density at this precursor concentration (Figure 5-4d).

Following a peak of cyanide production in the late exponential and early stationary phases, the cyanide concentration decreased during prolonged incubation (Figure 5-4a and Figure 5-4b). Some cyanogenic soil bacteria including *Pseudomonas* are known to convert cyanide into β-cyanoalanine during the late stationary and decay phase (Knowles, 1976). Moreover, *P. fluorescens* utilizes cyanide as a nitrogen source and converts it to ammonia (Kunz et al., 1992). Decomposition of cyanide by bacteria might have possibilities for overcoming the toxicity issues. Finally, sensitive process control might be required in an upscaled system, due to the relatively short duration of stable biogenic cyanide production.

Hydrocyanic acid (HCN) has a pK_a of 9.4 and the concentration of free cyanide (CN⁻) in solution is highly dependent on pH. Cyanidation, the process of leaching metals with cyanide, is typically carried out in alkaline conditions at pH over 10.5 (Akcil et al., 2015). In our experiments, the biogenic cyanide process was observed to occur between pH 7.3 and 9.4, below the pK_a of HCN, which is partly volatile in its associated form. A decreased free cyanide concentration was possibly a result of volatilization to HCN gas and its decomposition by *Pseudomonads* (Knowles, 1976). The increase of pH during gold bioleaching experiments, as shown in Figure 5-5, could favor the metal mobilization efficiency by increasing the chemical stability of CN⁻ in solution. Nevertheless, a biotechnological approach for gold recovery using biologically produced CN⁻ requires a balance between the chemical stability of the complexing/lixiviating agent and bacterial physiological requirements.

Table 5-1:Overview of recent bioleaching of copper and gold from printed circuit boards.

PCB type	Leaching mechanism	Process parameters			Efficiency (%)		Reference
		pH	T°C	Time (d)	Cu	Au	
Bioleaching of copper							
PCB (not specified)	Acidolysis and redoxolysis	2-3	37	15	53	-	Karwowska et al., 2014
Television	Acidolysis and redoxolysis	1.7	35	5	84	-	Bas et al., 2013
Personal computer	Acidolysis and redoxolysis	2.5	28	9	99	-	Wang et al., 2009
Bioleaching of gold							
Personal computer	Complexolysis	7-10	30	8	-	22.5	Natarajan and Ting, 2014
Personal computer	Complexolysis	7-10	25	5	-	8.2	Ruan et al., 2014
Bioleaching of copper and gold							
Mobile phone	Complexolysis	10	30	8	10.8	11.4	Chi et al., 2011
Personal computer	Complexolysis	7-9	30	7	not assayed	68.5	Brandl et al., 2008
Personal computer	Acidolysis, redoxolysis and complexolysis (two-step process)	1-2	23	7	98.4	44.6	This study
		7-9	30	2			

Chemical leaching of gold showed similarity with biogenic cyanide leaching in terms of the metal mobilization efficiency. The gold cyanidation reaction is as shown in equation (5). Stoichiometrically, the cyanide concentration from bacterial cultures should be enough to mobilize all the gold present in the material (22 µg/g PCB, Figure 5-5c). However, the relatively low efficiency indicates the consumption of cyanide by other agents, such as other noble metals or residual cu and other metals not leached during the first step.

$$4Au + 8CN^- + O_2 + 2H_2O \rightarrow 4Au(CN)_2^- + 4OH^- \qquad (5\text{-}5)$$

Base metals, e.g. copper, nickel, iron, and zinc form stable complexes with cyanide. Their presence at a high concentration would interfere with gold cyanidation. Preferential metal dissolution over gold may be due to two reasons: i) the prevalence of base metals consumes the free cyanide that would otherwise be available for gold complexation, and ii) more reactive metals in the residue form complexes with cyanide. Indeed, in reference to standard electro potential gold (Au^0/Au^+; $E^0 = -1.83$ V) is less reactive than other metals such as nickel (Ni^0/Ni^{2+}; $E^0 = -0.67$ V), iron (Fe^0/Fe^{2+}; $E^0 = -0.44$ V) and copper (Cu^0/Cu^{2+}; $E^0 = -0.34$ V) prevalent in PCB. Therefore, it is important that base metals are removed in the first step of the two-step process. In comparison to earlier studies, our results showed higher metal removal efficiencies for both metals, as shown in Table 2. To further optimize the metal mobilization from PCB, a wider spectrum of metals, including rare earth elements (REE) are recommended to be assayed in the characterization of the waste material and during the leaching steps.

5.4.3 Metal concentration of PCB

Characterization of the waste material displayed an abundance and a large variety of base and precious metals in PCB. The metal concentrations obtained in this study are in good agreement with those investigated with other researchers applying a similar methodology with regard to pre-treatment, particle size, and digestion method (Veit et al., 2006; Gurung et al., 2013). Variations in metal compositions of PCB may be attributed to different applied characterization methods, changing manufacturing and design techniques as well as technological innovations.

Copper is the predominant metal in PCB due to its high electrical conductivity. The variation of copper concentrations in various PCB stems from the type of the board (single layer or multi-layer), and year of manufacture (Yamane et al., 2011). Boards of all kinds have observably decreased in size and multi-layered boards are more widespread used (Marques, 2013). This results in a more compact product and an increase in copper concentration in printed circuit boards throughout the years (Ghosh et al., 2015). Gold, along with other precious metals, is used as a thin layer contact material in PCB due to its chemical stability. A significant amount of gold is concentrated on the components, due to the concentration difference between waste groups. PCB contain a 10-100 times higher concentration of gold and 30-40 times higher concentration of copper than natural ores (Akcil et al., 2015). Therefore, waste PCB can be considered as a promising secondary source of copper and gold.

Variation of nickel, lead and chromium concentrations in PCB may be attributed to advancing manufacturing and soldering technology. Mobile phone PCB, due to their compact size and design, have higher concentrations of relatively more valuable metals such as copper, nickel, chromium and gold, but lower concentrations of iron and lead (Shah et al., 2014). This could be a result of advancing board technologies and soldering techniques, as well as restrictions on the usage of certain substances, e.g. lead as per the restrictions dictated by a number of directives, such as the Restriction of Hazardous Substances Directive (RoHS Directive 2002/95/EC) by the European Commission. The high abundance of iron in the PCB is of further significance as it might suggest that the PCB itself could provide an energy source for iron-oxidizers for the bioleaching community. Further investigation is required to have a better understanding of the speciation of iron in the bioleaching medium, especially because leached iron can be recycled to the acidophilic bioleaching medium, where it can be an electron donor for iron-oxidizers.

5.5 Conclusions

Using a two-step approach, copper and gold are removed from PCB with an efficiency of 98.4% and 44.0%, respectively. The pre-growth strategy contributed to maintain optimal bioleaching conditions and cell viability. Separating bioleaching processes for base and precious metals with distinctive chemical properties enables an effective metal recovery strategy in terms of mobilization efficiency. Copper bioleaching with *Acidithiobacillus ferrivorans* and *Acidithiobacillus thiooxidans* was feasible at pH 1.0-1.6 and ambient temperature (23 ± 2°C). Gold bioleaching was effective with *Pseudomonas putida* and *Pseudomonas fluorescens* s but the gold mobilization efficiency by biogenic cyanide is lower compared to chemical cyanide. There is thus a need for further research aiming to enhance bacterial cyanide production, as well as increase the chemical stability of the free cyanide in solution.

Chapter 6.

Two-step leaching of valuable metals from discarded printed circuit boards, process kinetics, and optimization using response surface methodology

Abstract

Waste electrical and electronic equipment (WEEE) is an important secondary source of valuable metals. Particularly discarded printed circuit boards (PCB) contain high concentrations of valuable metals, varying greatly among the type of boards, the manufacture year, source device, and the production technology. Hydrometallurgical processing is an efficient way to selectively extract and subsequently recover metals from discarded high grade PCB. In this work, we propose a two-step process to extract copper (Cu) and gold (Au) from a high grade telecom server PCB. The boards contained 262.4 and 0.320 mg/g Cu and Au, respectively, which constituted the 98.1% of the total value. The metal extraction process was optimized using response surface methodology (RSM) by central composite design (CCD). The optimized process parameters showed that 3.92 M H_2SO_4, 3.93 M H_2O_2, 6.98% (w/v) pulp density and 3.7 hours contact time, and 0.038 M $CuSO_4$, 0.3 M $S_2O_3^{2-}$, 0.38 M NH_4OH, 10.76% pulp density (w/v) 6.73 hours were optimal for the maximal extraction of Cu and Au, respectively. At optimal conditions, 99.2% and 92.2% of Cu and Au, respectively, were extracted from the discarded PCB.

6.1 Introduction

Discarded printed circuit boards (PCB) are an important secondary source of valuable metals. All electrical and electronical equipment (EEE) contain PCB (Marques, 2013) of various size, type and composition (Duan et al., 2011). These materials are a complex mixture of metals, polymers and ceramics (Yamane et al., 2011). Low lifespan of electronic devices (Zhang et al., 2012a), perpetual innovation in electronics (Ongondo et al., 2015), and affordability of the devices (Wang et al., 2013) resulted in an unprecedented increase of waste electrical and electronic equipment (WEEE). In 2014, global WEEE generation was 41.8 million tons (Mt), of which 9.5, 7.0 and 6.0 Mt belonged to EU-28, USA and China, respectively (StEP, 2015), and is likely to increase to 50 Mt in 2018 (Baldé et al., 2015). The hazards associated with improper WEEE management comes twofold, **(1)** degradation of the environment (Song and Li, 2014) and **(2)** loss of valuable resources (Oguchi et al., 2013). Despite its potential toxicity, WEEE contains valuable materials that could be recovered to conserve primary resources coupled with economic benefits.

WEEE is a complex mixture of different materials in various concentrations. Modern devices encompass up to 60 elements, with an increase of complexity with various mixtures of compounds (Bloodworth, 2014). These elements go into the manufacture of microprocessors, circuit boards, displays, and permanent magnets in very complex alloys (Reck and Graedel, 2012). The composition of PCB after processing from a WEEE treatment plant is 38.1% ferrous metals, 16.5% non-ferrous metals, and 26.5% plastic, 18.9% others (Bigum et al., 2012). Precious metals are the main driver of recycling (Hagelüken, 2006), viz. gold (Au) has the highest recovery priority; followed by copper (Cu), palladium (Pd), aluminum (Al), tin (Sn), lead (Pb), platinum (Pt), nickel (Ni), zinc (Zn) and silver (Ag) (Wang and Gaustad, 2012) owing to the individual value and criticality of these metals, On the other hand, the intrinsic value of non-precious metals are increasing (Tanskanen, 2013) owing to decreasing concentration of precious metals, and elevated concentration of 'new' technology metals, such as the rare earth elements (REE) (Luda, 2011; Yang et al., 2011).

Several strategies were experimented to extract valuable metals from PCB through hydrometallurgical approaches. Hydrometallurgical metal recovery routes involve an oxidative leaching for the extraction of metals and a subsequent recovery and refining steps (Schlesinger et al., 2011). Metal extraction via oxidative acidic or alkaline leaching medium is an essential process in metal recovery from PCB (Ghosh et al., 2015). Leaching of metals from discarded PCB in various media including, hydrochloric acid (HCl) (Havlik et al., 2010; Lee et al., 2015), sulfuric acid (H_2SO_4) (Yang et al., 2011; Rocchetti et al., 2013), nitric acid (HNO_3) (Joda and Rashchi, 2012; Janyasuthiwong et al., 2016), often in addition of an oxidant such as hydrogen peroxide (H_2O_2) (Xiao et al., 2013), ferric iron (Fe^{3+}) (Li and Miller, 2007), chloride (Cl^-) (Kim et al., 2011a) and oxygen (O_2) (Dai and Breuer, 2013) have been been reported. Moreover, several novel leachants including thiosulfate ($S_2O_3^{2-}$) (Ha et al., 2010; Petter et al., 2014), thiourea (Jing-ying et al., 2012; Birloaga et al., 2014), and iodine (I_2) (Sahin et al., 2015; Serpe et al., 2015) were investigated for their effectiveness to recover valuable metals from waste PCB.

H_2SO_4 leaching is the most common method, often in combination with an oxidant, to recover base metals from their primary (Schlesinger et al., 2011) and secondary sources (Cui and Zhang, 2008), including discarded PCB. H_2O_2 is a strong oxidant (1.8 V), which is commonly used in combination with acids in order to enhance metal extraction yields. The oxidation reaction is exothermic and control of the temperature may be needed (Wang et al., 2016). Concentrations of lixiviant (H_2SO_4) and oxidant (H_2O_2) are the most influential factors affecting metal extraction from WEEE. Earlier studies investigated H_2SO_4 leaching of Cu from discarded PCB at non-optimized high reagent concentrations (Yang et al., 2011) and with addition of electro-generated oxidation agent (Ping et al., 2009), typically with high metal extraction efficiencies.

Precious metal leaching with $S_2O_3^{2-}$ is a non-toxic alternative to cyanidation for primary ores (Grosse et al., 2003) and secondary ores (Akcil et al., 2015). The Au leaching rates can be faster than conventional cyanidation, a lower interference from other cations is prevalent, a high yield we can be obtained, and a full scale operation can be more cost-effective than cyanidation (Abbruzzese et al., 1995; Aylmore and Muir, 2001). $S_2O_3^{2-}$ in presence of NH_4^+ and cupric ions (Cu^{2+}) leaches and complexes precious metals such as gold (Au) and silver (Ag). It allows the solubilization of Au as a stable anionic complex at alkaline or near neutral solutions (Abbruzzese et al., 1995) in which Cu^{2+} acts as a catalyst in the dissolution reaction (Akcil et al., 2015), and NH_4^+ as a stabilizing agent of the system (Senanayake, 2005) and thereby accelerate the anodic dissolution (Senanayake, 2004).

Leaching of metals from waste material is a complex multivariable process and dependent on a number of chemical and physical parameters and their reciprocal interaction. Factorial design of experiments (DoE) is a powerful tool to analyze the process parameters. Central composite design (CCD) with response surface methodology (RSM) is a robust statistical tool whereby multiple parameters and their interaction influence are involved in the selected response. RSM is increasingly used in hydrometallurgical processing of ores, in order to maximize the yield under optimal process parameters (Vegliò and Ubaldini, 2001; Azizi et al., 2012; Biswas et al., 2014). CCD with RSM simultaneously compute several involved factors at various levels and analyze the model for the relation between the various factors and their response (Niu et al., 2016).

In this study, we propose a multi-step leaching procedure for Cu and Au from discarded PCB. In the first stage, H_2SO_4 leaching of Cu in the presence of H_2O_2 was carried out. In the second step, Au was leached using $S_2O_3^{2-}$ as lixiviant with Cu^{2+} as catalyzer in ammoniac medium ($CuSO_4$–NH_4OH–$Na_2S_2O_3$). In the first step, four operational parameters namely sulfuric acid H_2SO_4 concentration, hydrogen peroxide H_2O_2 concentration, contact time (h) and pulp density (%, w/v) were optimized in order to maximize Cu yield. Similarly, in the second leaching process, five operational parameters, namely $S_2O_3^{2-}$ concentration, copper sulfate ($CuSO_4$) concentration, ammonium hydroxide (NH_4OH) concentration, contact time (h) and pulp density

(%, w/v) were optimized in order to maximize Au yield. Finally, the model predictions were compared to experimental data in confirmatory tests.

6.2 Materials and methods

6.2.1 Source, preparation and characterization of the printed circuit boards (PCB)

Discarded printed circuit boards (PCB) from telecom server devices and desktop computers were collected from SIMS Recycling in Eindhoven, and the IT department of UNESCO-IHE, the Netherlands. Desktop computer PCB were used for metal concentration comparison. The samples were prepared as described earlier in Chapter 4. After the milling of each sample, the ground materials were sieved to particle sizes of <500 μm to be used in leaching experiments. The boards are shown in Figure 6-1. The characterization of waste PCB was done by a modified

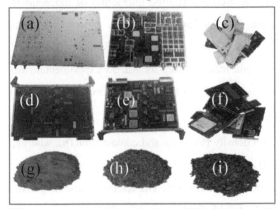

EPA method (3052, 1996: Microwave assisted acid digestion of siliceous and organically based matrices), as described earlier in Chapter 4. Ground PCB was acid-digested in a laboratory microwave unit (CEM, MARS, USA) in mixture of nitro-hydrochloric acid (HNO_3 and HCl, 1:3 by volume). The digestates were serially diluted for metal

Figure 6-1: Discarded telecom PCB samples of Telecom boards (T1) front (a), (b), (c), (d), (e), (f); after crushing and sieving to particle size <500 μm (j), 500 – 1600 μm (h) and 1600 – 2500 μm (i).

analysis. The metals values were taken for the London metal exchange on a 1-year basis between 1 August 2015 and 1 August 2016 (LME, 2016).

6.2.2 Design of experiments (DoE) and optimization

Design of experiments (DoE) is a statistical tool to the analyze and optimize the independent process variables for maximum efficiency by evaluating the interactive effects of operational parameters and minimizing the number of experiments (Liang et al., 2013). Central composite design (CCD) was used to study effects of variation of parameters together with response surface methodology (RSM). For the maximal metal yield from the discarded boards, RSM provided statistically supported DoE and mathematical methods for model design, analysis of interaction of the process parameters for the two-step leaching of metals, and the computation of optimal conditions for the parameters.

The total number of experiments was calculated using a factorial design (2k) as given below:

$$N = 2^k + 2k + n \qquad\qquad (6\text{-}1)$$

where (N) is the total number of experiments, k is the number of process variables, and n is the number of replicates, i.e. the center points.

Two CCD were designed, each with two level factorial points, axial points and central points for the optimization of the two-step Cu and Au leaching process (Table 6-1). Evaluated responses were Cu yield (mg Cu/g PCB) and Au yield (mg Au/100 mg PCB). Levels of coded and assigned variables, and each factor was studied at 5 different levels (-α, -1, 0, 1, +α) in the design. MINITAB 17 (Minitab Inc, USA) was used analyze the interaction of independent process variables for the RSM study. In the first Cu leaching step, 31 experimental runs including 24 full factorial experiments (Runs 1–24), and 7 center points (Runs 25–31) with 4 factors and 5 levels were designed. In the second leaching step for Au, 53 experimental runs including 32 full factorial experiment (Runs 1–32) and 17 center points in axis (Runs 43–52) with 5 factors and 5 levels were designed.

For Cu leaching, H_2SO_4, (M), H_2O_2 (M), pulp density (%, w/v), and contact time (h), whereas for Au leaching, $CuSO_4$ (M), $S_2O_3^{2-}$ (M), NH_4OH (M), pulp density (%, w/v), and contact time (h) were considered as variable operational process parameters. The selected value ranges for both H_2SO_4 and H_2O_2 concentrations were 2.0 – 6.0 M, the PD was 4.0 – 10.0 % (w/v), contact time was 2.0 – 6.0 hours, based on earlier studies (Birloaga et al., 2013; Kamran et al., 2013; Zhou et al., 2013). Similarly, for Au leaching the selected range for $CuSO_4$ concentration was 0.025 – 0.05 M, for both $S_2O_3^{2-}$ and NH_4OH concentrations were 0.2 – 0.4 M, for the pulp density was 1.0 – 10.0 % (w/v and for time was 4.0 – 8.0 hours. The parameters and their ranges were selected in line with the earlier published reports conducted with similar discarded materials (Breuer and Jeffrey, 2000; Ha et al., 2010; Petter et al., 2014), and the concertation of the targeted metals in the discarded PCB given in Chapter 4.

Table 6-1: CCD of parameters from low to high levels for copper (Cu) and gold (Au) leaching.

Parameter	Unit	Code	Range and levels				
			- α	- 1	0	+ 1	+ α
Copper leaching yield							
H_2SO_4	(M)	A	0.0	2.0	4.0	6.0	8.0
H_2O_2	(M)	B	0.0	2.0	4.0	6.0	8.0
PD	(%, w/v)	C	1.0	4.0	7.0	10.0	13.0
Time	(h)	D	0.0	2.0	4.0	6.0	8.0
Gold leaching yield							
$CuSO_4$	(M)	A	0.0	0.025	0.0375	0.05	0.75
$S_2O_3^{2-}$	(M)	B	0.0	0.2	0.3	0.4	0.6
NH_4OH	(M)	C	0.0	0.2	0.3	0.4	0.6
PD	(%, w/v)	D	2.0	4.0	6.0	8.0	12.0
Time	(h)	E	0.0	2.5	3.75	5.0	7.5

6.2.3 Confirmatory leaching experiments

Additional sets of confirmation tests were carried out in order to confirm the validity and reliability of RSM model predictions. The operational parameters of confirmation experiments were selected as per the CCD design and the results of the RSM study (optimal and -1 values), plus control tests. Discarded PCB were put in contact with the corresponding concentrations of leachant and oxidant solutions in 100 mL Erlenmeyer flasks were closed with a septum screw cap. The flasks were agitated at approximately 150 rpm at constant ambient temperature (23 ± 2°C). Periodic metal analyses were carried out by syringes with 5 mL volume. The analytes were filtered (Whatmann 0.45 μm G/C), centrifuged at 1000 rpm for 10 min, washed three times and dried at 105°C for 1 h for subsequent analysis and diluted for metal measurements. At the end of the first leaching step, the residues were collected and characterized for their total metal content as per the procedure described in 6.2.1. In the following Au leaching experiments, residues from the preceding leaching step were washed with ultrapure water and put in contact with the corresponding concentrations of leachant, oxidant and catalyzer solutions in 100 mL

Erlenmeyer flasks and agitated at approximately 150 rpm at constant ambient temperature (23 ± 2°C). Each test was carried out in duplicate.

6.2.4 Analytical methods

The metal analyses were performed by inductively coupled plasma mass spectrometer (ICP-MS) (X series 2, Thermo Scientific, USA) and inductively coupled plasma optical emission spectrometer (ICP-OES) (Perkin Elmer, Optima 8300, USA) as described in Chapter 4.2. Analytical grade chemicals and ultrapure (MiliQ) water were used. Alternatively, periodic metal measurements were carried out by atomic absorption spectrophotometer (AAS-F) (Varian 200, USA). The readings for the following wavelengths (nm) were recorded for Cu 324.752, and Au 267.595. The analyzing blanks and calibration standards, were performed to establish the reproducibility of the data standards and assure the analytical quality standards. The statistical analysis of the results was carried out by MINITAB 17 and Microsoft Excel 2016.

6.3 Results

6.3.1 Metal concentrations and intrinsic value of the discarded PCB

The concentrations of copper (Cu) and gold (Au), of the telecom and desktop boards are shown in Table 6-2. Telecom boards had a higher concentration of both metals over regular computer board (PCB). The intrinsic value of Au was higher in telecom boards, owing to its high concentration in telecom boards (15.2 factor). The intrinsic value of Cu and Au was 51.6 and 41.3 respectively for desktop PCB, on the other hand, the intrinsic value of Au and Cu in Telecom PCB was 85.8 and 11.3, respectively. In both cases, these two metals constituted the most value of other metals, 91.9% for desktop boards, and 97.1% of the total value.

6.3.2 Optimization of the metal extraction process using response surface methodology (RSM)

6.3.2.1 *Optimization and desirability function*

Response surface methodology (RSM) using central composite design (CCD) for copper (Cu) and gold (Au) leaching determined the optimized process parameters for the maximal yield of the two metals. The desirability functions for Cu and Au leaching are given in Figure 6-2. The optimal value for the maximized response of multiple influencing factors i.e. leachant, oxidant and catalyzer concentrations, pulp density, and contact time were considered to directly influence the yield of metals extracted from discarded PCB.

Table 6-2: Metal concentrations (mg/g PCB) of the discarded printed circuit boards.

Metals	Cu (mg/g PCB)	Au (µg/g PCB)
Telecom board	262.4 ± 16.6	320.7 ± 17.1
Intrinsic value (%)	11.3	85.8
Desktop computer	176.7 ± 23.6	21 ± 3.2
Intrinsic value (%)	51.6	41.3

A desirability function was incorporated to achieve the maximum metal extraction yield for each leaching step. For the ranges selected, viz., 2.0 – 6.0 M H_2SO_4, 2.0 – 6.0 M H_2O_2, 1.0 – 10.0% (w/v) pulp density and 0.0 – 6.0 h contact time, the desirability function for the maximum for Cu yield is given in Figure 2a. Similarly, for Au leaching 0.025 – 0.05 M $CuSO_4$, 0.2 – 0.4 M $S_2O_3^{2-}$, 0.2 – 0.4 M NH_4OH, 4.0 – 8.0% pulp density (w/v), and 2.5 – 5.0 h contact time, the desirability function is given in Figure 6-2b. These values were used to perform confirmatory tests under the optimized conditions in order to ascertain the reliability of the model.

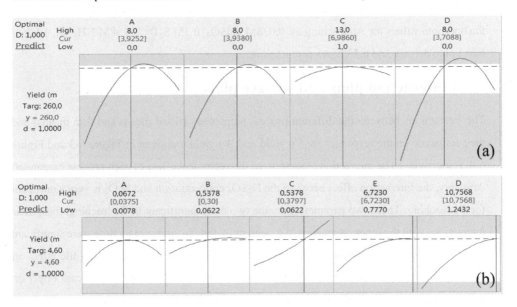

Figure 6-2: Desirability function for (a) Cu for (A) H_2SO_4 (M), (B) H_2O_2 (M), (C) pulp density (%), w/v, and (D) contact time (h) and (b) gold leaching and for (A) $CuSO_4$ (M), (B) $S_2O_3^{2-}$(M), (C) NH_4OH (M), (D) pulp density (%, w/v), and (E) contact time.

Regression equation of Cu yield is shown below:

$$Y = -243 + 61.8A + 58.3B + 5.6C + 77.2D - 6.81A^2 - 6.68B^2 - 0.692C^2 - 8.46D^2 \tag{6-2}$$
$$+ 0.27AB + 0.57AC + 0.61AD + 0.30BC + 1.08BD + 0.45CD$$

Where Y represents the Cu yield in (mg/g PCB), (A) the H_2SO_4 concentration in M, (B) the H_2O_2 concentration in M, (C) the pulp density in % (w/v), and (D) the contact time in hours. Within the range of low and high values of the parameters studied, a non-linear optimization protocol was followed according to the Monte-Carlo optimization procedure. Equation (6-2) was solved using the Monte-Carlo optimization technique and the results gave the optimum values for Cu leaching as: 3.92 M H_2SO_4, 3.93 M H_2O_2, 6.98%, w/v) pulp density and 3.7 h contact time. Similarly, the regression equation for Au leaching is shown below in equation (6-2).

$$Y = -1.30 + 155.1A + 10.16B - 12.57C - 0.304D + 0.464E - 1844A^2 - 9.16B^2 +$$
$$6.75C^2 - 0.0928D^2 - 0.0364E^2 - 58.3AB - 25.0AC - 2.55AD + 3.93AE + 0.4BC \tag{6-3}$$
$$+ 1.703BD - 1.078BE + 1.402CD + 0.714CE + 0.0455DE$$

Where Y represents Au yield in (mg/100mg PCB), (A) $CuSO_4$ concentration in M, (B) $S_2O_3^{2-}$ concentration in M, (C) NH_4OH concentration in M, (D) contact time (E) pulp density. Equation (6-2) was solved using the Monte-Carlo optimization technique and the results gave the optimum values for Au leaching as: 0.038M $CuSO_4$, 0.3M $S_2O_3^{2-}$, 0.38M NH_4OH, 10.76% (w/v) pulp density and 6.73 hours contact time.

6.3.3 Interaction plots and surface plots

The interactions between the different process parameters (fitted means and data means) and their influence on the responses for Cu yield and Au yield are given in Figure 6-3 and Figure 6-4, respectively. The interaction plots showed the reciprocal effect of each process parameter. Arguably, the interaction effect between the H_2SO_4 concentration and H_2O_2 is very significant for Cu leaching. These two parameters are the two most significant in Cu leaching, thus, their interactions with the other parameters, e.g. with pulp density, and contact time were significant as well. The interaction between pulp density and time, on the other hand, had little effect on Cu leaching. An increase of pulp density from 2.0 to 6.0 (w/v) had an insignificant effect on the final Cu yield. Similarly, the increase of contact time from 1.0 to 6.0 hours did not a significant effect on the final yield. The variation in those two parameters had a relative low influence on the Cu yield, as shown in Figure 6-3a).

The effect of the interaction between the leachant, the oxidant and the catalyzer concentrations had a direct effect on Au yield, however the effect of time and pulp density, was not very prevalent. These results showed similarity to leaching of Cu, where physical parameters played a less significant role than the concentrations of the reagents. $S_2O_3^{2-}$ was arguably the most significant parameter for the yield. Therefore, its interaction with the other parameters significantly affected the final leeching efficiency. The interaction of $S_2O_3^{2-}$ and NH_4OH concentrations had a very significant effect on Au leaching (Figure 6-4). Thus, it can be concluded the effect of the chemical concentration and process parameters are less significant than the effect of collateral chemical concentrations. These findings were confirmed by the results from 3-D surface plots, which gave complete and statistically significant information regarding the effect of each factor and the interaction between factors on the response. The 3-D response surface plots expressing the metals yields (0 levels of CCD in Table 7-1 were carried for the process parameters for Cu leaching (Figure 6-3) and Au leaching (Figure 6-4), which showed the effect of a selected process variable, while the other two process parameters were kept constant.

When the concentrations of H_2SO_4 and H_2O_2 were increased, from $-\alpha$ value to $+\alpha$, the Cu yield increased to higher values and eventually reached maximum extraction yield of 262 mg Cu /g PCB at 3.92 M H_2SO_4 and 3.93 M H_2O_2 concentration at constant optimal values of pulp density and contact time. Similarly, a statistically significant interaction between H_2SO_4 and PD was observed, when H_2O_2 was kept at constant optimal values, and the effect of these parameters on the dissolution concentration had a positive correlation up to the optimal process parameters. Exceeding the optimal concentration, the effect was however, found to be negative, as the increase of the concentration over optimal value toward the maximal value affected the yield negatively. Besides, increasing concentration of H_2SO_4 and contact time at constant, H_2O_2 had a positive effect on the yield.

Concerning the Au yield, the concentrations $S_2O_3^{2-}$ and $CuSO_4$ were increased, from $-\alpha$ value toward $+\alpha$, the Cu yield increased to higher values and eventually to maximum yield of 320 mg Au/100 g PCB at and 0.038 M $CuSO_4$, 0.3 M $S_2O_3^{2-}$, 0.38 M NH_4OH, concentration at constant optimal pulp density and contact time values. NH_4^+ concentration has a less prevalent effect than the other two chemical, and its interaction with $S_2O_3^{2-}$ had a small effect on Au yield (Figure 6-4).

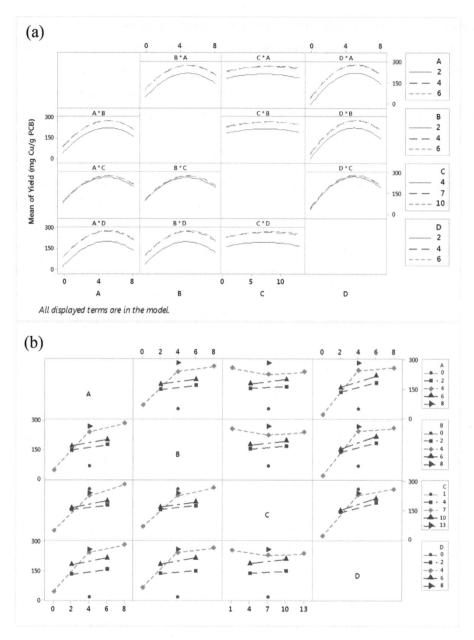

Figure 6-3: Interaction plot for Cu leaching process variable fitted means (a) and data means (b) for (A) H$_2$SO$_4$ (M), (B) H$_2$O$_2$ (M), (C) pulp density (%), w/v, and (D) contact time (h).

(a)

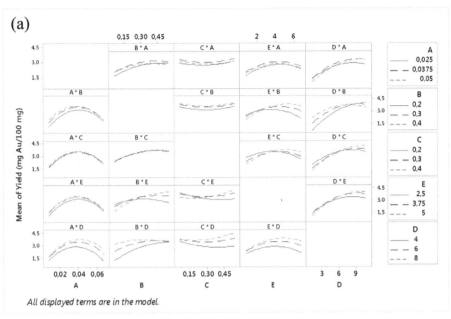

All displayed terms are in the model.

(b)

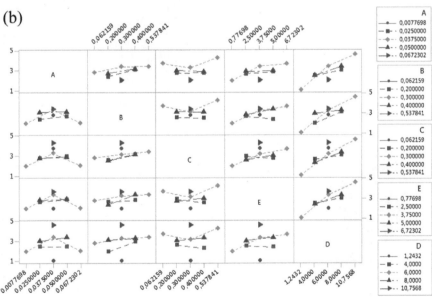

Figure 6-4: Interaction plot for Au leaching process variable fitted means (a) and data means (b)(A) CuSO₄ (M), (B) S₂O₃²⁻ (M), (C) NH₄OH (M), (D) pulp density (%, w/v), and (E) contact time.

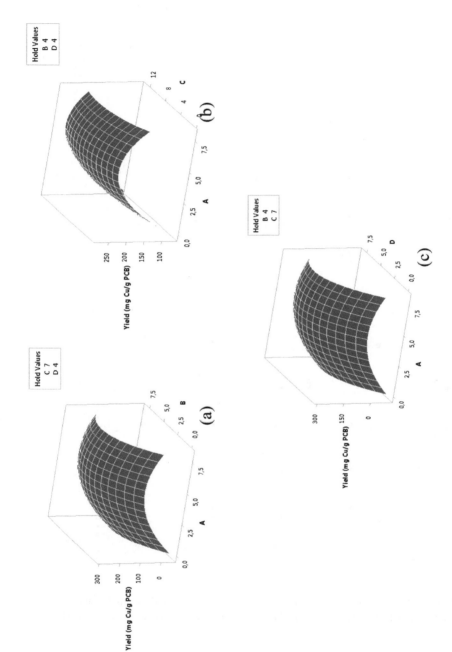

Figure 6-5: Surface plots of Cu extraction yield versus for H_2SO_4 (M), H_2SO_4 (M), (PD) (%, w/v), and contact time, thiosulfate ($S_2O_3^{2-}$) and (a) pulp density (PD) (%, w/v) (b) time and (c) ammonium thiosulfate (NH_4OH).

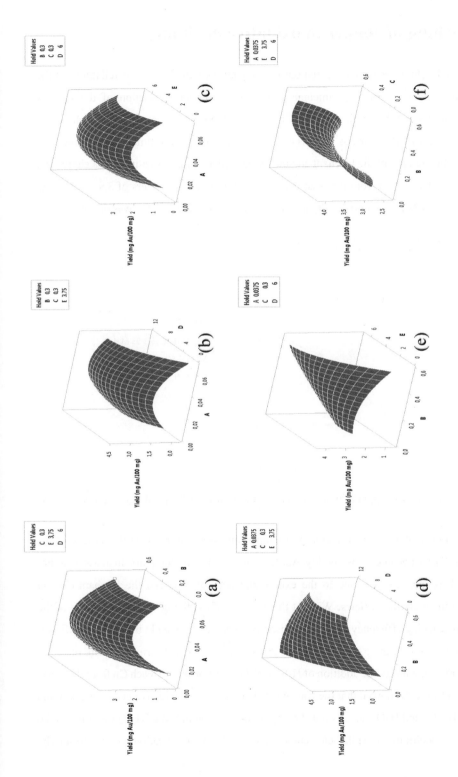

Figure 6-6: Surface plots of gold extraction yield versus for (A) thiosulfate ($S_2O_3^{2-}$) and (a) pulp density (PD) (%, w/v) (b) time and (c) ammonium thiosulfate (NH_4OH).

6.4 Leaching of copper in oxidative medium

Confirmatory leaching tests were carried out to compare the validity and the reliability of the model predictions with the experimental data. The reciprocal interaction of the process parameters and the yield at optimal process parameters were investigated. The dissolution of Cu from discarded PCB at various process parameters as a function of time is demonstrated in Figure 6-7. The increase of the leachant and the oxidant concentrations resulted in a significant increase of Cu dissolution from the waste material. At optimal parameters of 3.92 M H_2SO_4, 3.93 M H_2O_2, 6.98% (w/v) pulp density and 4 h contact time, 99.2% of Cu was dissolved.

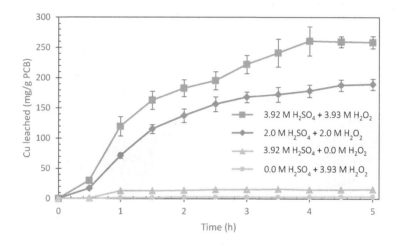

Figure 6-7: Leaching of copper (Cu) as a function of time with H_2SO_4 and H_2O_2 at 6.98 % (v/w) pulp density.

During the confirmatory tests an increase in temperature was observed, as the surface of the Erlenmeyer flasks became considerably warm, however, the temperature increase was not measured. This increase was due to the exothermic reaction between the leachant-oxidant mixture and the metallic particles. Without the addition of oxidizing agents or heating of the solution, the leaching efficiency of Cu was observed to be very low and virtually insignificant. Without the addition of the oxidant H_2O_2, 6.52% of Cu was leached form the waste material. On the other hand, without the addition of H_2SO_4, H_2O_2 was not able to leach Cu from the waste material. Control tests showed that the leaching of Cu was due to oxidative acidic attack of the mixture of H_2SO_4 and H_2O_2 (results not shown). Without the addition of H_2O_2, the solution did not reach the conditions to oxidize elemental copper (Cu^0) to cupric cations (Cu^{2+}). Similarly, without the addition of the leachant H_2SO_4, Cu^0 was not dissolved into the leachate solution. In negative control tests, no Cu leaching was observed (results not shown).

6.5 Leaching of gold in ammoniacal thiosulfate medium

Similar to Cu leaching, confirmatory leaching tests were also carried out to validate the RSM model predictions for Au leaching. Under optimal conditions of 0.38 M $CuSO_4$, 0.3 M $S_2O_3^{2-}$, 0.38 M NH_4OH, 10.76% (w/v) pulp density, 58.2% and 71.4% of the Au leached at the end of the 2nd and the 3rd hour. Eventually 92.2% of the Au was dissolved in 7 hours (Figure 8).

The varying $S_2O_3^{2-}$ concentration had a significant impact on Au leaching yield. In the confirmatory tests, the concentration of $S_2O_3^{2-}$ was varied in the range 0.1 to 0.3 M, while the NH_4OH and $CuSO_4$ concentrations were kept constant at their optimal values. Lower $S_2O_3^{2-}$ concentrations than the optimal value resulted in a lower Au yield. 51.8% and 77.0%, of the total Au was leached respectively at 0.1 and 0.2 M of $S_2O_3^{2-}$. In control tests without $S_2O_3^{2-}$ addition, no Au leaching was detected (results not shown), which indicated that the $S_2O_3^{2-}$ concentration was the main factor in Au leaching.

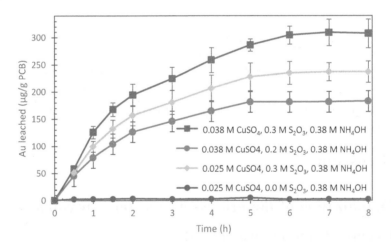

Figure 6-8: Leaching of gold (Au) with $S_2O_3^{2-}$ catalyzed by $CuSO_4$ in NH_4OH medium as a function of time, and at 10.76 % (v/w) pulp density.

Varying the $CuSO_4$ concentration, had an effect on the final yield but not on the kinetics of the leaching reaction. Lower $CuSO_4$ concertation (0.025 M) resulted in lower Au (59.4%) extraction yield compared to optimal values, when the other operational parameters are kept constant (Figure 7). Without $CuSO_4$ addition, dissolution took place but only 21.4% of Au was dissolved (results not shown). The rate was found to decrease with time as the $CuSO_4$ concentration decreased.

The relative error between the measured data of the confirmation experiments and the calculated results of the models was 97.2% for Cu and 97.3% for Au, respectively. The regression of the model predictions and the experimental data is given in Figure 6-9. Since the value predicted by the model was within the 95% confidence interval, this can be taken as confirmation of the suitability of the regression model for predictive purposes. Therefore, the RSM approach and CCD developed in this study provides reliable predictive data for the Cu and Au leaching. When scaled up a sensitiy analysis would be required.

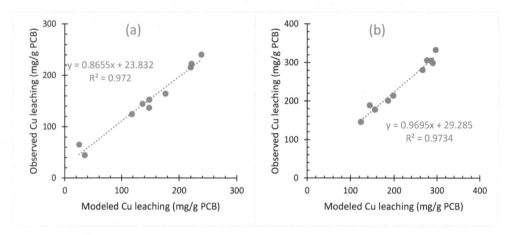

Figure 6-9: Regression analysis of the model prediction with the measured data for (a) Cu leaching and (b) gold leaching.

6.6 Discussion

6.6.1 Optimization of the process parameters for copper and gold leaching from discarded PCB

Selective metal recovery from such a complex anthropogenic polymetallic secondary resource requires a novel approach for the development of an efficient metal recovery process. Diversity of metals and complexity of metal–metal and metal–non-metal associations inflict specific challenges (Tuncuk et al., 2012). Hydrometallurgical processes enable relatively low capital costs, no hazardous gas emission, operational selectivity for small scale applications, and are propitious alternatives to conventional pyrometallurgical processes for metal recovery. In this chapter, a two-step approach to sequentially leach copper (Cu) and gold (Au) from a high grade PCB is proposed.

Response surface methods (RSM) is a collection of mathematical and statistical techniques used for developing, improving and optimizing the processes (DoE) particularly in complex hydrometallurgical metal extraction process where multiple independent variables are involved. At the initial stage of process development, RSM revealed the optimal conditions and interaction of the process variables where maximal yield of desired products can be achieved. It produces statistically-validated predictive models, interaction plots of the independent variables and, response surface maps that point the way to pinnacles of the process performance.

The results suggest that the reagent concentrations are more significant factors than the other process variables in hydrometallurgical extraction of valuable metals from discarded PCB at high pulp densities (5 - 10%, w/v). This applied to both to the leaching of Cu and Au. The optimum concentrations of the leachant, H_2SO_4 and H_2O_2 for Cu and $S_2O_3^{2-}$ for Au, played a dominant role during leaching of these metals. In addition, there were significant interactions between H_2SO_4 and H_2O_2 for Cu and between $S_2O_3^{2-}$ and $CuSO_4$ for Au. The effect of PD and time were rather less prominent for both Cu and Au extraction yield. Furthermore, the results indicated that the interaction between leachants and the other independent variables was statistically insignificant. As a result, the optimum metal yield from discarded PCB did not depend on the level of PD and time. This does, however, not translate into the insignificance of these parameters, but in a much smaller influence of them on the main yield.

The analysis of variance (ANOVA) showed that the response surface quadratic model was significant at F value of 5.04 and a P values of <0.0001. The statistical significance of the model was also confirmed by the determination of coefficient ($R^2 = 0.911$) which indicated that only 8.1% of the variations were not explained by the model and this also means that 91% of the variations were explained by the independent variables. The model prediction and the optimized values were in good agreement. One exception was the effect of NH_4OH concentration on Au yield, the model predicted a positive correlation between the Au extraction yield and NH_4OH concentration above the optimal value (Figure 6-4). However, in the confirmatory tests it had insignificant effect (Figure 6-8).

6.6.2 **Characterization of discarded PCB**

Telecom devices PCB showed differences from consumer electronics PCB such as regular personal computers in terms of metals concentrations (Table 6-2). Typically, they have higher concentrations of Cu and Au, along with other valuable metals. The economic value of Cu and Au constituted 98.1% of the total value of the metals measured. Thus, the main economic motivation for discarded PCB recycling is the selective and efficient recovery of Cu and Au. These two metals react similarly in hydrometallurgical reaction and a prevalent competition is usually observed. Thus, an innovative strategy is required for the processing of polymetallic high grade secondary source for metal recovery.

The Cu fraction is overlaid by laminate layers in multilayer PCB, the most widely used type of boards, which inhibit the contact between the leaching medium and Cu (Havlik et al., 2010). Thus, liberation of the components and the Cu layer is required to the largest extent possible, which is typically carried out by mechanical crushing of the discarded boards in specialized comminution equipment. Au is found predominantly on the central processing unit (CPU), up to 96% of this elements is found on these central units (Luda, 2011). In metal assay studies, the relatively low Au content of the CPU-removed desktop PCB (Table 1) confirmed these findings. The rest of the Au is found in the contact layers of these boards, typically built intact with the board matrix and is most efficiently separated by a selective leaching process. PCB from telecom boards contain Au as a contact material at the backside of their top layer (Figure 6-1) did not contain visible conventional CPU.

The metal concentrations varied largely per particle sizes (Table 1), which is of practical significance for the selection and design of a pre-treatment step. Yoo et al. (2009) argues that the conventional crushers are not well-suited to cut and crush such a friable and brittle material. Also, sieving to smaller than 0.5 mm leads to the loss of the material, presumably on the sieve and thus accumulation of certain metals can occur on particular mesh sizes. Physical pre-treatment of discarded PCB leads to significant material losses (Chancerel et al., 2011). Physical separation processes can be used ahead of hydrometallurgical processes at the expense of metal losses. In addition to the optimization of the leaching and purification steps, it is recommended to also optimize the physical pre-treatment, perhaps with more advanced equipment designed to process such material and particle size distributions.

6.6.3 **Leaching of Cu from the discarded PCB**

It is of practical significance for the development of a metal recovery process to acknowledge that, most metals in WEEE are found in their elemental form, or as alloys, distinctive to primary ores (Dai and Breuer, 2013). In general, an oxidative leaching process is required for the effective extraction of base metals. Concentrations of H_2SO_4 and H_2O_2 were the main process variables that control the extraction of Cu from the waste material, as was predicted by the optimization model and confirmed in the confirmatory experiments. The higher dosage of H_2SO_4 can release more Cu through proton attack and H_2O_2 provides an oxidative medium to facilitate the leaching of Cu into the solution.

$$Cu^0 + 2H^+ \rightarrow Cu^{2+} + H_2 \qquad \Delta G^0 = 65.5 \text{ kJ/mol} \qquad (6\text{-}4)$$

$$Cu^0 + 1/2\, O_2 + 2H^+ \rightarrow Cu^{2+} + H_2O \qquad \Delta G^0 = \text{-}171.63 \text{ kJ/mol} \qquad (6\text{-}5)$$

$$Cu^0 + H_2O_2 + H_2SO_4 \rightarrow Cu^{2+} + SO_4^{2-} + H_2O \qquad \Delta G^0 = \text{-}329.7 \text{ kJ/mol} \qquad (6\text{-}6)$$

Leaching of Cu from PCB in the absence and the presence of the oxidant in acidic sulfate media showed a difference in metal mobilization. In the H_2SO_4 leaching system, addition of H_2O_2 significantly increased the Cu dissolution rate. In fact, without the addition of the oxidant, Cu cannot be effectively leached into solution. A very limited Cu extraction (2%) in the absence of the oxidant is consistent with the thermodynamic information, as in equations (6-4), (6-5) and (6-6). Moreover, the presence of H_2O_2 as a strong oxidant is likely to prevent the reduction or precipitation of the oxidized metal species in the solution (Xiao et al., 2013).

6.6.4 **Leaching of Au from the discarded telecom PCB**

The dissolution of Au from discarded PCB in NH_4^+ medium is a process mediated by the concentration of $S_2O_3^{2-}$ in the presence of Cu^{2+} ions as the oxidant. Cu^{2+} ions create the thermodynamically possible conditions for the leaching of gold into the solution. The major role of NH_4^+ in the thiosulfate system is to stabilize Cu^{2+} ions. The Au leaching solution includes the thermodynamically stable cupric-tetraammine complex as a result of the mixture of the $CuSO_4$ and NH_3 in aqueous medium as shown below:

$$Cu^{2+} + NH_3 \leftrightarrow Cu(NH_3)_4^{2+} \qquad (6\text{-}7)$$

The cupric-tetraammine ($Cu(NH_3)_4^{2+}$) complex is a thermodynamically stable species which enhances the stability region of Cu(II)–Cu(I), preventing the reduction of Cu into solid compounds. Thus, the Cu^{2+} concentration of the Au leaching solution is an important factor for the thermodynamical stability of the solution. The CCD model corroborated a high correlation of $CuSO_4$ concentration and Au yield (Figure 4b and Figure 6). In our confirmatory tests, the initial rate of gold extraction -within the first two hours- is enhanced with increasing $CuSO_4$ concentration, however, not significant above a certain concertation (Figure 8). Varying the Cu concentration, from 0.01 to 0.03 M $CuSO_4$, did not influence the predominant Cu species in the reaction system at the same Eh/pH conditions. Moreover, a very low Au dissolution efficiency was observed in the tests in the absence of Cu^{2+} (Figure 6-8).

Au dissolution with $S_2O_3^{2-}$ occurs in presence of cupric-tetraammine as the oxidant which forms stable $Au(S_2O_3)_2^{3-}$ and complexes, as shown below in Equation (6-8):

$$Au + 5S_2O_3^{2-} + Cu(NH_3)_4^{2+} \rightarrow Au(S_2O_3^{2-})_2^{3-} + 4NH_3 + Cu(S_2O_3)_3^{5-} \qquad (6\text{-}8)$$

At the optimal conditions, the rate of Au dissolution was rapid during the first two hours of the reaction, then the extraction reached the steady state (Fig. 8). The maximum gold recovery obtained was 96.22% after 7 h of leaching. $S_2O_3^{2-}$ had a direct influence on gold leaching, as predicted by CCD model and verified in the confirmatory tests. Such a result is not surprising since the dissolution reaction Equation (6-8) is likely to be enhanced at increased thiosulfate concentrations. In terms of process development, a balance between leach kinetics and high thiosulfate consumption is required; an optimum value appears to be 0.38 M from a high-grade copper removed PCB.

Varying the NH_4^+ content had an insignificant effect in our experimental conditions, while the kinetic curve showed the same shape. Earlier studies reported decreasing gold recovery with increasing the NH_4^+ concentration due to disturbed thermodynamic stability of $Cu(NH_3)_4^{2+}$. However, these studies reported concentrations much higher than the process variables optimized in this work. Leaching of gold from PCB in this concentration range of target metal, leachant and oxidant is found to be feasible. Conclusively the two-step approach, in which a potential competition between metals is prevented can be regarded as an efficient strategy to leach valuable metals from the waste material. Moreover, a two-step step extraction approach leads to two separate leachate solutions in different media which might be of practical importance to achieve selective recovery of metals from discarded PCB material.

6.7 Conclusions

Anthropogenic secondary raw materials, dissimilar to primary ores, are very complex and require a novel metal recovery approach. In this study, the procedure for a two-step hydrometallurgical route for the extraction of valuable metals from a high grade WEEE, containing 260 mg/g Cu and 0.320 mg/g Au was developed. A two-step Cu and Au extraction procedure was designed and optimized by RSM using the CCD technique. The model accurately predicted the yields under various operational conditions with high coefficients of determination between the response and the process variables. In confirmatory tests, 99.2% and 92.2% of Cu and Au was extracted, respectively, under the optimized conditions. Oxidative acid leaching for the extraction of Cu from discarded PCB is a kinetically fast and efficient technology. A mixture of an oxidant and a leachant is required for efficient Cu extraction from the waste material. The leaching of Au in ammonical thiosulfate solutions was found to be chemically controlled.

Chapter 7.

Selective recovery of copper from the leachate solution by sulfide precipitation and electrowinning

Abstract

This chapter aimed to selectively recover valuable metals from a real acidic leachate solution. The solution originated from an earlier study where the metals were leached from crushed discarded computer printed circuit boards (PCB) via biotechnological routes. The leachate solution contained high concentrations of copper (Cu), iron (Fe), aluminum (Al), and minor levels of zinc (Zn). The recovery efficiencies, process kinetics, the properties of the final products were studied for two selected techniques, namely sulfidic precipitation and electrowinning. In precipitation experiments with sulfide solutions, copper was selectively precipitated at low pH, with a Cu:S molar ratio of 1:1, without requirement of any pretreatment. 83% and 95% of copper was precipitated at 35 mM and 100 mM of sulfide concentrations, respectively. The average particle size distribution was 152.1 (d.nm.)and zeta potential was -30.3 mV for precipitates at 35 mM sulfide concentration. In electrowinning tests, 97.8% of copper was recovered at 100 mA/cm^2 current density in 60 minutes. Current density correlated positively with the process kinetics. The Faraday yields and specific energy consumption were 76.3%, 69.2% and 55.7%; 6.75, 7.43, and 9.24 Wh/g of copper recovered respectively, at current densities 50, 100 and 200 mA. The results showed the feasibility of both techniques for selective recovery of metals.

7.1 Introduction

Waste electrical and electronic equipment (WEEE) is an important secondary source of metals. Printed circuit boards (PCB) are particularly rich in metals including copper (Cu), iron (Fe), aluminum (Al), zinc (Zn), nickel (Ni), silver (Ag), gold (Au), platinum (Pt), and palladium (Pd) (Wang and Gaustad, 2012). The metal concentrations depend on the board type, the manufacturer, year of manufacture, and assaying method (Ghosh et al., 2015). Many researchers reported hydrometallurgical (Tuncuk et al., 2012; Xiao et al., 2013) and biohydrometallurgical processes (Lee and Pandey, 2012; Chen et al., 2015), to recover metals from discarded electronics. These routes typically involve a leaching step, which results in a multi-metal leachate solution with high concentrations of oxidized metallic cations, leached from metal-rich waste material.

In our chapter 5, we bioleached copper along with other base metals from discarded PCB with an iron-and sulfur-oxidizer bacterial co-culture (Fowler et al., 2015). The constructed mixture consisted of pure cultures of *Acidithiobacillus ferrivorans* (DSM 17398),and *Aciditihiobacillus thiooxidans* (DSM 9463). The process resulted in a leachate solution that is rich in metallic cations. In a consecutive recovery step, we aim to selectively recover metals from leachate solution containing high concentrations of metals. In this direction, two techniques, namely sulfidic precipitation, and electrowinning were considered to recover metals from the LS. Selective recovery of Cu from the leachate solution, the primary metal found in discarded PCB was of priority. Moreover, in the bioleaching process Fe was used as a source of energy by the iron-oxidizing bacteria. In a broader resource recovery perspective, recirculation of Fe back to the bioleaching process is a specific strategy. Thus, selective recovery of copper, and leaving Fe in the leachate solution is the primary objective of this work.

Sulfidic precipitation gained interest particularly applications of effluent treatment processes such as acid mine drainage (AMD) treatment (Sahinkaya et al., 2009) and industrial hydrometallurgical processes (Reis et al., 2013). Sulfidic precipitation advantages include lower solubility of the precipitates relative to hydroxide precipitation, high dewaterability of metal sulfide sludges, highly selective metal removal, high degree of metal removal even at low pH, fast reaction rates, good settling properties (Tokuda et al., 2008; Lewis, 2010). A generic precipitation reaction of copper with sulfide is given below in equation (7-1).

$$Cu^{2+} + S^{2-} \rightarrow CuS_{(s)} \qquad (7\text{-}1)$$

Sulfidic precipitation is a complex phenomenon resulting from the induction of supersaturation conditions (Sampaio et al., 2010). Precipitation occurs through three stages: **(i)** nucleation, **(ii)** crystal growth, and **(iii)** flocculation. Solubility product constants can be predicted (at equilibrium conditions) for thermodynamic calculations (Veeken et al., 2003).

Electrowinning (EW) process, on the other hand, entails a direct electrical current from an external source from the anode to the cathode through an electrolyte. Metals in the electrolyte deposits onto the cathodes using the energy provided by the electrical current to drive the reduction of the cations to their elemental form metal. The final products of EW process are pure metal at the cathode, and formation of gas at the anode.

The main cathodic process of cuprous and cupric electrodeposition reactions in sulfate media are given in equations (7-2) and (7-3) below:

$$Cu^{2+} + 2e^- \rightarrow Cu^0 \ (E^0 = 0.337\,V) \qquad (7\text{-}2)$$

$$Cu^+ + e^- \rightarrow Cu^0 \ (E^0 = 0.506 \ V) \tag{7-3}$$

It is also accompanied by secondary reactions:

$$Cu^{2+} + e^- \rightarrow Cu^+ \ (E^0 = 0.167 \ V) \tag{7-4}$$

The reactions can be generalized to other metallic cations as well:

$$Ni^{2+} + 2e^- \rightarrow Cu^+ \ (E^0 = -0.25 \ V) \tag{7-5}$$

$$Fe^{2+} + 2e^- \rightarrow Cu^+ \ (E^0 = -0.44 \ V) \tag{7-6}$$

$$Zn^{2+} + 2e^- \rightarrow Cu^+ \ (E^0 = -0.76 \ V) \tag{7-7}$$

$$Al^{3+} + 3e^- \rightarrow Cu^+ \ (E^0 = -1.66 \ V) \tag{7-8}$$

Electrowinning cells utilize a design comprising cathodes and anodes. Current is passed through the electrolyte and moves the cations toward the cathode. As the metal in solution is deposited onto the cathode, the liquid film adjacent to the cathode becomes depleted in metal (Dimitrov, 2016). The current intensity determines the rate of metal deposition, and should be at a level which cations diffuse to the cathode.

This study describes an electrowinning cell to be used in novel waste (bio)processing for metal recovery. Approximately one thirds (4.5 million tons) of copper is produced by electrowinning (EW) annually (Schlesinger et al., 2011). Production of copper using this technology continues to increase due to the growth of leaching as a metal recovery process technology for copper (Mukongo et al., 2009). It has advantages of high process control, high final product purity and being environmentally friendly with electrons being the only reagents of the process. The process parameters depend on the characterization of the leachate solution, the metals of interest and their speciation (Kordosky, 2002). The application areas of EW spans from primary sources (Kamran et al., 2013), to acid mine drainage (Gorgievski et al., 2009), and to recovery of metals from secondary sources (Kasper et al., 2011). These new application fields require novel process design strategies in order to selectively recover metals with high purity.

The fundamentals into the aspects of both sulfide precipitation (Lewis and Van Hille, 2006) and electrodeposition (Panda and Das, 2001) are well understood and main process parameters are explained. However, they are restricted to very low concentrations, which are of limited value in most process-based metal recovery applications. In this work, we intend to apply sulfide precipitation and electrowinning to a real WEEE acidic leachate solution and compare their efficiency in terms of reaction kinetics, selectivity towards copper, and investigate the structural and chemical properties of the final products.

The main objective of this work is to **selectively recover copper** from a real metal-rich leachate liquor solution resulting from a bioleaching process. In this direction, a real leachate solution was used to study the process parameters and identify potential bottlenecks for practical process development. In this direction, a set of experiments was conducted using the real WEEE acidic leachate solution. The specific objectives are to **(i)** investigate the effect of various process variables **(ii)** compare the efficiency of the two selected techniques **(iii)** study the properties of the final products of the recovery processes **(iv)** evaluate the efficiency of the two techniques with a view on their applicability on a larger scale and integrability with the preceding leaching processes.

7.2 Materials and methods

7.2.1 The leachate solution and its characterization

In this chapter, a leachate solution from bioleaching of discarded printed circuit boards (PCB) from computers was used. The composition of the board was (%, *w/w*) 17.6 Cu, 5.2 Fe, 3.6 Al, 0.5 Zn, 0.2 Ni. Bioleaching process was described in detail in our previous work (Fowler et al., 2015) in which additional ferrous iron (Fe^{2+}) was supplemented as an energy source for the acidophilic microorganisms. Copper was leached with high efficiency (>98.4%). Prior to this work, the leachate solution was filter-sterilized and stored at 4°C. The leachate solution was characterized by measuring cations; Cu^{2+}, total Fe and Fe^{2+}, Al^{3+}, Ni^{2+}, Zn^{2+}; the anions; SO_4^{2-}, Cl^- and NO_3^-, total organic carbon (TOC) and physical parameters such as temperature (T°C), conductivity (mS), and pH. In order to investigate the various and compare it to a fully chemical solution a synthetic leachate solution was prepared. The synthetic solution was prepared by chemical reagents $CuSO_4 \cdot 5H_2O$, $Fe(II)SO_4 \cdot 7H_2O$, $AlK(SO_4)_2 \cdot 12H_2O$, and $ZnSO_4 \cdot 7H_2O$ at concentration corresponding to the real leachate solution. The characterization results of the leachate solution and the composition of the synthetic solution are given in Table 7-1.

Table 7-1: Characterization of the leachate solution and the composition of the synthetic solution.

Leachate solution		Synthetic solution	
Metals	**mg/L**	**Metals**	**mg/L**
Cu^{2+}	2187 ± 142	Cu^{2+}	2180
Total Fe	1713 ± 65	Fe^{2+}	1720
Fe^{2+}/Fe^{3+} ratio	0.94 ± 0.01	Al^{3+}	290
Al^{3+}	290.1 ± 34	Zn^{2+}	50
Zn^{2+}	51.6 ± 6		
Total organic matter　**mg/L**			
TOC	39.1 ± 1.6	-	-
Anions	**mg/L**		
SO_4^{2-}	3693 ± 321	SO_4^{2-}	8330
Cl^-	21.0 ± 2.6		
NO_3^-	68.5 ± 64		
Physicochemical parameters			
pH	1.05 ± 0.02	pH	1.05 ± 0.01
ORP (mV)	460 ± 6	ORP (mV)	580 ± 4
T°C	23 ± 2	T°C	23 ± 2
Conductivity (mS/cm)	56 ± 5	Conductivity (mS/cm)	72 ± 4

7.2.2 Sulfidic precipitation experiments

In precipitation experiments, selective recovery of Cu^{2+} over other cations at acidic low pH was aimed. Aqueous precipitant solutions were prepared with sodium sulfide nonahydrate ($Na_2S \cdot 9H_2O$) at various sulfide (S^{2-}) concentrations (1-500 mM). The metal-rich leachate solution and the precipitant were mixed at equal volumes of 0.5 mL, in 1.5 mL reaction tubes. The mixture was sampled at various time intervals (1-100 min). The supernatant was analyzed for particle size distribution (PSD) and zeta potential measurements. Following the settling of the precipitates, solid liquid separation was carried out by centrifuging at 10,000 rpm for 10 min. The supernatant was collected, diluted and passed through a membrane filter (0.22 μm), acidified (1%, *v/v* concentrated HNO_3) for metal measurements. The reaction was modeled using Visual MINTEQ Version 3.1 to predicted theoretical thermodynamic calculations.

7.2.3 Electrowinning experiments

As dissolved species of metals in the leachate solution have widely disparate reactivity levels, and in particular iron has multiple oxidation states with different reactivity levels, it was not apparent whether it was possible to achieve some selectivity in electrodepositing. Hence, electrodeposition experiments were used to investigate the behavior firstly of synthetic solution (SS) along with the LS.

7.2.3.1 *Voltammetry tests*

Linear and cyclic sweep voltammetry tests were carried out to obtain information on the current potential behavior of the metal cations in the solutions. A calomel reference electrode (REF421, Radiometer Analytical, France) in a saturated KCl solution (0.199V vs. SHE) was used for the voltammetry tests. Linear sweep voltammetry was performed with a potential scan of 25 mV/s from -1.2 V to 1.6 V (vs. Ag/AgCl) with the leachate solution in an electrolytic cell of 0.5 L volume with two platinum electrodes as anode and cathode stirred at 1000 rpm and the calomel reference electrode connected to a potentiometer (Voltalab, Model PST006, Radiometer Analytical, France). The results were recorded by a software (Voltamaster 4, Radiometer Analytical, France). The cell potential versus current density was recorded in 0.25 s intervals using the software's data acquisition system.

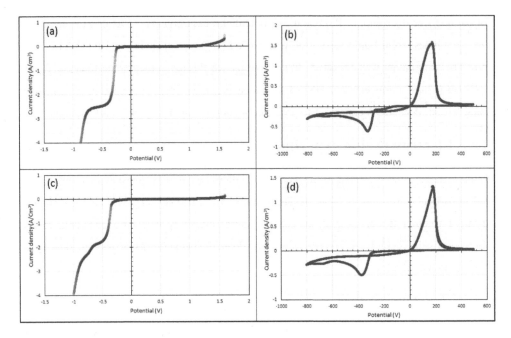

Figure 7-1: Linear (a), (c) and cyclic sweep (b), (d) voltammetry tests of the synthetic solution (SS) and the real leachate solution respectively.

7.2.3.2 *Electrodeposition tests*

In electrodeposition experiments, an electrolytic cell with working volume 100 mL was employed, with a platinum anode and one carbon felt cathode with leachate solution as the electrolyte. A potentiostat/galvanostat and a DC Triple power supply unit (HAMEG, HM8040-3, Switzerland) were connected to an anode of pure platinum and a pitch-based filamentous carbon fiber felt cathode. The fibers were approximately 10 μm in diameter as observed in the electron microscope. The carbon felt cathodes were prepared at 1×1×4 cm (L×W×H), covered with tape and submerged at approximately 0.5 cm below the surface of the electrolyte. The electrolyte was agitated with a magnetic stirrer. The schematic of the electrolytic cell is given in Figure 7-2.

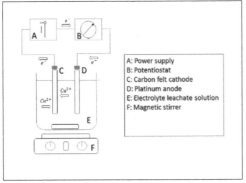

A: Power supply
B: Potentiostat
C: Carbon felt cathode
D: Platinum anode
E: Electrolyte leachate solution
F: Magnetic stirrer

Figure 7-2: Schematic electrolytic cell design.

The current was set at values between 50 and 200 mA in order to allow the selective deposition. This corresponded to 2.8, 3.9, 6.1 V of electric potential for each current, respectively, for 50 mA, 100 mA and 200 mA. All experiments were performed over a period of 180 min. The electrolyte leachate solution and the synthetic solution were sampled for the metal concentrations at 10 minutes' intervals for the entire experimental time. Changes in the current were monitored from the display of the potentiometer. All experiments were performed in triplicate at $23 \pm 2°C$ in open atmosphere.

7.2.3.3 *Current (Faradic) efficiency and power consumption*

The power requirement is calculated by multiplying current and voltage, as shown below in equation (7-9).

$$P = V.I \tag{7-9}$$

Where P is power consumption in watts and V is voltage in volts and I is the current in amperes.

Energy consumed per mass of copper is calculated by multiplying power consumption by time consumed divided by the mass deposited on the cathode as shown below:

$$E_c = P.t/m \tag{7-10}$$

Where E_c is the energy consumption in Wh/g and m is the mass of copper deposited in gram.

The specific current (Faradaic) efficiency is calculated according to the formula below, derived from Faraday's law of electrodeposition as shown below:

$$\eta = \frac{n.F.m}{M.I.t} \tag{7-11}$$

Where η denotes efficiency, m is the weight of copper recovered on the cathode in g, M is the molar mass of copper (63.5 g/mol), I is the current in ampere, t is the electrodeposition time in seconds, n is the number of electrons involved in the reaction and F is the Faraday constant, i.e. 96,485 C/mol.

7.2.4 **Properties of the final products**

The morphology, visual characterization and the chemical composition of the precipitates and the cathodic deposits were examined by scanning electron microscopy (SEM) equipped with energy dispersive X-ray analyzer (EDS) (JSM-6010LA, JEOL, Japan). The precipitates were collected with a plastic fine bore Pasteur pipette and placed on a brass pin covered with graphite. Particle size distribution (PSD) and zeta potential analyses were conducted by a zeta sizer (Malvern, Nano ZS, UK) with laser light scattering technique.

The composition, morphology and metallic depositions were observed over time during the processes. Cathodic deposits were stored under nitrogen atmosphere prior to their analysis in order to prevent oxidation. Metal concentrations of the leachate solutions were periodically measured during both recovery processes, in order to investigate the process kinetics and removal efficiency. The solutions were visually observed for changes in physical parameters and as an indication of concentration. The carbon felt cathodes were collected and stored for mass balance analyses. The surface morphology and the chemical composition of the cathodic deposits were investigated by SEM-EDS (JSM-6010LA, JEOL, Japan).

7.2.5 **Analytical methods**

Cu, Fe, Al, Zn were analyzed at wavelengths (nm) 324.8, 259.4, 394.4, 202.6 respectively by inductively coupled plasma optical emission spectrophotometry (ICP-OES) (Perkin Elmer Optima 8300, USA) equipped with a solid-state SCD detector and low-flow argon nebulizer (Meinhard K1, USA). Alternatively, periodic metal measurements were carried out by atomic absorption spectrophotometer (AAS-F) (Agilent 200, Varian, USA) at the above-mentioned wavelengths for the individual metals. Multi-element calibration standards were of analytical grade in 5% HNO_3 matrix (Perkin Elmer, France). Ferrous iron (Fe^{2+}) was measured by a modified o-phenanthroline method as described by Herrera et al. (1989). Anion measurements were carried out by ICS-4000 Dionex ion chromatograph (Thermo Scientific, USA). The instrument was equipped with a dual pump (DP) and an eluent generator module, which generated a high purity KOH eluent from deionised water using a EGC-KOH cartridge, an IC Cube module, which included an injection valve, a degas cartridge, column heater, guard column (IonPac AG15, 0.4 × 50 mm), separator column (IonPac AS15, 0.4 × 250 mm). A capillary conductivity detector (CD) was used for detection. The samples were injected via an auto-sampler (Dionex AS-DV, Thermo Scientific, USA). Total organic carbon (TOC) was analyzed in a total organic carbon analyzer (TOC-L, Shimadzu, Japan) by non-purgeable organic carbon method as explained by Oturan et al., (2015). pH was measured by Ag/AgCl reference electrode (SenTix 21, WTW, Germany) connected to a pH meter (691 Metrohm pH meter, Germany). Conductivity was measured by an electrode (CDC566T, Radiometer Analytical, France) connected to a conductivity meter (CDM230, Radiometer Analytical, France).

7.3 Results

7.3.1 Sulfidic precipitation of metals

Removal efficiencies of the metals from the leachate solution at various precipitant concentrations are given at Figure 7-3. Copper was observably isolated from the LS in our experimental conditions. The reaction of metals and aqueous sulfide was almost instantaneous, resulting in the formation of insoluble metal–sulfide complexes. Thus, the effect of time was found to play negligible role (results not shown) and the results were calculated according to the samples taken at t=1 min. At sulfide concentrations of 35 mM and lower, no odorous H_2S formation was detected.

The solution was turbid as in a slurry, as the crystallization occurred in various particles sizes, including very small ones. The removal of metals increased with the increasing sulfide concentration. Copper was removed with 5.0%, 16.5%, 83.1% 95.2%, 99.3% and 99.9% efficiencies at sulfide concentrations of 1, 10, 35, 100, 200 and 500 mM, respectively (Figure 7-3). 10 mM sulfide concentration resulted in the rapid formation of a CuS precipitate with good settling characteristics, although 63.2% of the copper remained in solution. At 35 mM of sulfide concentration 83.4% of Cu was removed from the solution. At a concentration of 500 mM, the metals are removed from the solution at high efficiencies, i.e. 99.9% of Cu, 96.4% of Fe, 97.1% of Zn and Al with 74.5%.

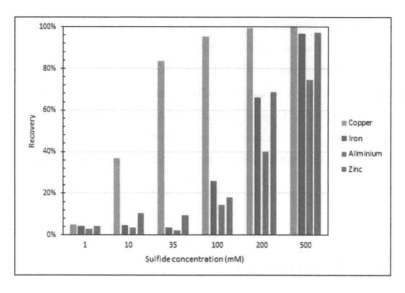

Figure 7-3: Metal removal from the leachate solution by sulfidic precipitation at various concentrations.

7.3.2 Zeta potential and particle size distribution

Zeta potential and PSD measurements results are given in detail in Table 7-2. Results show that the Zeta potentials were (in mV) -15.5 (± 0.12), -25.8 (± 0.2), -28.0 (± 0.14), -30.3 (± 0.23), -15.9 (± 0.12), -14.8 (± 0.08) respectively, for sulfide concentrations of (in ppm) 1, 10, 35, 100, 200 and 500. The zeta potential became less negative with an increase in the metal:sulfide (Me:S^{2-}) molar ratio up to the saturation concentration of 35 ppm. At concentration of 100 ppm, the zeta potential showed decreasing trend, where at 200 ppm was at its lowest. the zeta potential obtained was substantially more positive in the presence of excess sulfide (over 35 mM) than those produced in the absence of excess sulfide. PSD showed a normal distribution in all experimental conditions, with a notably small particle size at 35 mM (152.1 d.nm). The attenuator was between 7 and 10 for concentration, showing indicating highly turbid solutions. At 100mM, where the excess sulfide is lowest, the solution showed highest attenuator. The zeta potential results showed constituency with particle size and attenuator data Table 7-2.

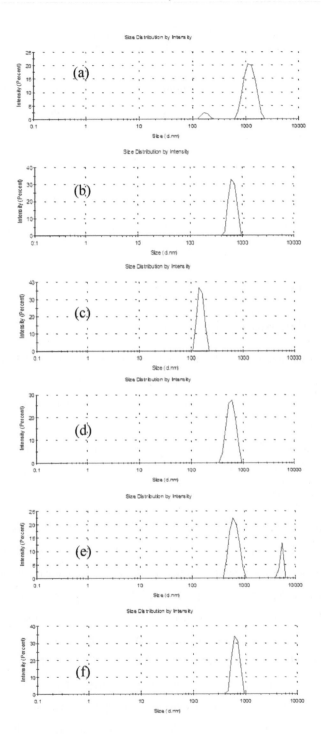

Figure 7-4: Particle size distribution of precipitates at sulfide concentration of (in ppm) 1 (a), 10 (b), 35 (c), 100 (d), 200 (e) and 500 (f).

Table 7-2: Zeta-potential (Zp), particle size distribution (PSD), attenuator, zeta average (Z-avrg.), % intensity, and standard deviation of the measurements (Stdev) measurements of the precipitates.

S^{2-} conc. (mM)	Zp (mv)	Stdev (mv)	Attenuator	In range	Z-avrg (d.nm.)	Stdev (d.nm)	% intensity
1	-15.5	0.12	7	73.7	1216	294.2	93.8
10	-25.8	0.2	7	75	659.5	102.8	100
35	-30.3	0.14	8	75.8	152.1	20.57	100
100	-18.0	0.23	10	66.5	594.4	110.8	100
200	-15.9	0.12	7	81.8	655.5	129.2	83.7
500	-14.8	0.08	6	87.6	655.5	95.86	100

7.3.3 Electrodeposition of metals

Results of the linear and cyclic voltammetry tests are given at Figure 7-1. Various cathodic deposition potentials in voltammetry results suggest that a number of intermediate complexes are formed and a difference in reduction potentials is prevalent. These are the minimum cathodic potentials of the cations, Cu^{2+}, Fe^{2+}, Fe^{3+}, Zn^{2+} and Al^{3+} in the leachate solution for the electrowinning process. The results of the voltammetry test with the synthetic solution showed similarity to the ones of the leachate solution with slight difference in the shape of the linear voltammetry curve (Figure 7-1).

Difference in potentials enabled selective deposition of the metals. Copper was selectively deposited on the cathode at the selected current range between 50 mA and 200 mA, showing various kinetic properties. Current density had a positive correlation with the deposition rate of the metals. Selective deposition of copper showed dependency on time and current, as shown in Figure 7-6. At current densities higher than 50 mA, zinc and iron co-deposition was significant. At 100 mA current density, 12% and 20% of iron and zinc, respectively, co-deposited at the cathode. At 200 mA current density, 16% and 14% of iron and zinc co-deposited. It should be noted, however, that the significance of these results are proportional with the concentrations of the metals in solution. Aluminum deposition was negligible in our experimental conditions. There was little difference between the leachate solution and synthetic solution, besides the kinetic properties of metal electrodeposition in synthetic solution was slightly higher (Figure 7-6).

Figure 7-5: Voltammetry tests (a), electrowinning experiments, carbon felt cathode after (c) and before deposition (d).

Despite an increasing interest and published works in recovery of metals from waste material using microbial routes (Wang et al., 2009; Shah et al., 2014), the final leachate solutions are seldom extensively characterized. This particularly applies to the organic content of the solution which is the final liquid product of bioprocessing of waste material. In this work, we detected a considerable amount of total organic content (39.1 mg/L, Table 7-1) in the leachate solution, which is not reported anywhere else, to the best knowledge of the author.

In terms of metal selectivity, no significant difference between the leachate solution and the synthetic leachate was observed, except for the slight difference of iron co-deposition, particularly at 100 and 200 mA current densities. At 100 mA and 200 mA, 88.2% and 94.4 and 84.2% and 88.4%, respectively, of the initial iron stayed in the solution at the end of the electrowinning process. The removal ratio was in line with the ferrous iron ratio of the total iron concentration between the real leachate (96%) and the synthetic solution (100%).

7.3.4 Energy consumption and current (Faradaic) efficiency

The electrowinning process was evaluated in terms of specific energy consumption, current (Faradaic) efficiency as a function of current used, reaction time and metal recovered. The power consumption and current efficiency was calculated for copper recovery using the equations (7-9), (7-10), and (7-11) given above in 7.2.3.3. Energy consumption was 6.74, 7.43, 9.24 Wh/g of deposited Cu, respectively, for 50 mA, 100 mA and 200 mA current densities. An increase in power consumption is observed with increasing current density. Current efficiency

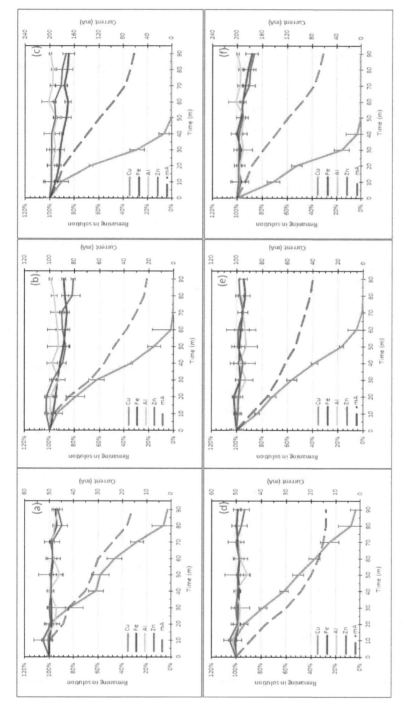

Figure 7-6: Recovery of metals from at 50 mA, 100 mA, and 200 mA current densities, respectively, from the leachate solution (LS) (a), (b) and (c) and the synthetic solution (SS) (d), (e) and (f) by electrodeposition.

for copper deposition was calculated as 76.3%, 69.2% and 55.2% at 50 mA, 100 mA and 200 mA current densities, respectively. The results are summarized in Table 7-3.

Table 7-3: Electrowinning power consumption and current efficiency calculations for the leachate solution.

Parameter	Unit	50 mA	100 mA	200 mA
Power	(W)	0.14	0.39	1.22
Current	(A)	0.05	0.1	0.2
Voltage	(V)	2.8	3.9	6.1
Energy consumption	(Wh g/L)	3.09	4.76	9.24
Mass	(g)	0.075	0.081	0.088
Molar mass	(g mol/L)	63.5	63.5	63.5
Current	(A)	0.05	0.1	0.2
Time	(s)	6000	3600	2400
Faraday constant	(C mol/L)	96,485	96,485	96,485
Number of electrons	(1 mol^{-1})	2	2	2
Current efficiency	(%)	0.76	0.69	0.55

7.3.5 Final products

The sulfidic precipitates and the cathodic deposits showed differences in morphology, chemical composition and physical appearance, as can be seen in Figure 7-7. SEM-EDS analysis revealed a porous morphology of the sulfidic precipitates and aggregation of particles without crystal forms. The color of precipitates depended on the reagent concentrations where brown for low concentrations; and a blue/black floc at higher concentrations. Copper and sulfide formed stoichiometric bounds according to their abundance, forming covellite mineral form, by a Cu:S ratio of 1:1. The concentration of copper and sulfide was 32.7 ± 1.2 and 32.8 ± 1.0 (%, by mass) as per the results of point sweep analysis by EDS. Small amount of iron was also observed in one-point analysis (Nr.2, Figure 7-7a) which indicated minor co-precipitation of this element. However, iron was not detected in other point EDS analyses.

Cathodic deposits from electrowinning experiments were in earth brown color, typical of elemental copper. EDS analysis on the deposits showed copper with a significant amount of oxygen, showing the presence of cupric oxide (CuO). This is due to deposition of very thin layer of copper on the carbon cathode and its spontaneous reaction with the atmospheric oxygen. Deposition was concentrated on the surface of the outer fibers and inclined toward the one side of the cathode, giving an overall heterogeneous distribution. Deposition onto the central fibers was strenuous, as they are randomly oriented and the excited copper ions preferentially approached the outer layer rather than the inner fibers.

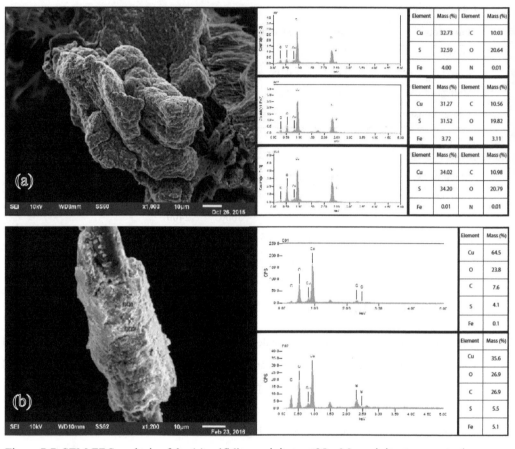

Element	Mass (%)	Element	Mass (%)
Cu	32.73	C	10.03
S	32.59	O	20.64
Fe	4.00	N	0.01

Element	Mass (%)	Element	Mass (%)
Cu	31.27	C	10.56
S	31.52	O	19.82
Fe	3.72	N	3.11

Element	Mass (%)	Element	Mass (%)
Cu	34.02	C	10.98
S	34.20	O	20.79
Fe	0.01	N	0.01

Element	Mass (%)
Cu	64.5
O	23.8
C	7.6
S	4.1
Fe	0.1

Element	Mass (%)
Cu	35.6
O	26.9
C	26.9
S	5.5
Fe	5.1

Figure 7-7: SEM-EDS analysis of the (a) sulfidic precipitate at 35 mM precipitant concentration and (b) cathodic deposition at 100 mA current density.

7.4 Discussion

7.4.1 Selective recovery of metals from the leachate solution

When multi-metal leachate solutions are concerned, it is possible to selectively recover and isolate them from the solution, using their various chemical differences. Many of these metals react readily with sulfide to form solid sulfide phases that have different solubility products. The key reactant involved is S^{2-}. On the other hand, dissolved cations have different reduction potentials of which can used to selectively deposit elemental metals by regulating the current.

In this study the aim was to target the (1) solubility constant differences of metals at given pH, and (2) difference in electrochemical reactivity. An order of selectively towards individual metals were observed towards individual metals was achieved in both recovery techniques.

In sulfidic precipitation tests, copper was selectively recovered, before the other cations in the solution, in our experimental conditions. In aqueous solutions of sparingly soluble these are expressed in terms of the solubility products (Van Hille et al., 2005). CuS has a solubility constant of $K_{sp} = 6 \times 10^{-37}$ at, much lower than those of iron (8×10^{-19}), Al (1×10^{-22}), and Zn (1×10^{-23}). Thus, Cu precipitated primarily, forming insoluble salts with S^{2-} ions, at a CuS ratio of 1:1, in form of covellite as confirmed by the EDS measurements. As the concentration of the precipitant increases other metal cations sequentially precipitated in order of (Fe> Ni> Zn). This follows the order of solubility of the respective sulfide salts of the metals. It should be noted that the addition of sulfide to the solution increases the pH according to the reaction below. By modifying the pH, it is possible achieve selectivity towards single metals. Cu at pH 1.0-1.5, Zn at pH 4.5-5.0 and Ni at pH 6.5-7.0 (Tokuda et al., 2008). An increased pH enabled metals to precipitate in their sulfide forms.

$$S^{2-} + H_2O \leftrightarrow HS^- + OH^- \tag{7-12}$$

From a thermodynamic point of view, only the metals with a higher electrochemical potential will co-deposit with Cu in electrodeposition tests. Kinetically, the reactions readily occurred, as opposed to sulfidic precipitation reactions using different sulfide sources such as H_2S (Tokuda et al., 2008). There was not any difference between the metals as the precipitation reactions occurred instantaneously.

Due to relative low reactivity of Cu, it was easily electrodeposited on the carbon cathode with high purity (65%) before the other metals. The behavior of other metal species present in the electrolyte can be understood with reference to their position in the reactivity series (Bebelis et al., 2013). Those species with a more positive reduction potential than Cu can plate on the cathode, while those with more negative potential remained in solution. Thus, selective deposition of copper was possible; due to its standard potentials difference. As can be seen from the reactions (7-13), (7-14), (7-15) below, in combination with (7-2), copper is much less reactive than the other ions in the solution.

$$Al^{3+} + 3e^- \rightarrow Al^0 \ (E^0 \ = \ -1.66 \ V) \hspace{2cm} (7\text{-}13)$$

$$Zn^{2+} + 2e^- \rightarrow Zn^0 \ (E^0 \ = \ -0.76 \ V) \hspace{2cm} (7\text{-}14)$$

$$Fe^{2+} + 2e^- \rightarrow Fe^0 \ (E^0 \ = \ -0.44 \ V) \hspace{2cm} (7\text{-}15)$$

The increase of intensity had an effect on the reaction rate but not the deposition of other cations. The current decreased during the electrodeposition reactions as with the decreasing conductivity. The small amount of iron co-deposited with copper, as can be seen at Figure 7-7. It could be attributed to the occurrence of the ferric ions along with ferrous ions in the solution (Das and Gopala Krishna, 1996). Selectivity toward copper was achieved and a competition between iron and copper was not observed, largely due to speciation of iron. Typically, a solvent-extraction step is employed so as to isolate one metal over the other, and prevent co-deposition (Kordosky, 2002; Dreisinger, 2006). With the leachate solution, where iron is found predominantly in ferrous form, copper was easily recovered, without the requirement of an additional step. At, only 4%, 9% and 12% of iron was co-deposited with copper at 50, 100 and 200 mA current densities (Figure 7-6). In its ferrous form, iron is much less electronegative than its ferric form which resulted in the selective deposition of copper over iron.

Effect of time was negligible in sulfide precipitation experiments as the reactions were instantaneous (results not shown). Formation of the precipitates with small sizes, however, resulted in a turbid slurry and the sedimentation of the precipitates was not affected by time. Formation of the miniscule particles hinted that the scaling up of the technology could be difficult task. Selective and sequential deposition proved feasible by alternating the current in electrowinning tests. Iron speciation is a crucial parameter in selective recovery of copper in and iron-and copper-rich leachate solution, as the reactivity level of ferrous iron ions is much lower than those of ferric ions, and it mediated the selective deposition of copper on the cathode in our experimental conditions. Additionally, the current density and the process kinetics were in positive correlation and increased current density improved the reaction rate. However, there is a balance between the current density and the process efficiency as the faradaic efficiency is reduced with the increasing current density. As the current intensity increased the current used to deposit copper on the electrode is higher.

7.4.2 **Energy consumption and process efficiency**

From an engineering point of view, selection of the most efficient current density with a high selectivity toward the targeted metal is desired. Faradaic current efficiencies for recovery of copper were relatively high at experimented current densities. In practice, a current efficiency above 60% is acceptable (Schlesinger et al., 2011), as it can be argued that in such complex solutions the current is consumed by many accompanying redox reactions, specifically the reduction of the reactive cations. A competing reaction to the deposition of copper is the reduction of Fe^{3+} to Fe^{2+} at the cathode. This explains the decreasing effect on efficiency of copper deposition, as some of the current is used for the reduction of ferric to ferrous (Das and Gopala Krishna, 1996). In addition, open atmosphere testing might be another reason of current consumption (Xiao et al., 2013), as the atmospheric oxygen could lead to oxidation of cuprous (Cu^+) to cupric (Cu^{2+}). Lastly, another reason could be that the relatively high current density for a dilute copper solution caused the copper to deposit as amorphous, porous copper fibers, which shortened electrode distance, decreased the cell voltage and efficiency.

The specific energy consumption was comparable to the findings of Gorgievski et al. (2009) who used similar experimental conditions; however higher than the findings of Vegliò et al., (2003) and Panda and Das (2001), who used synthetic solutions. The main reason for relatively high energy consumption is related to the usage of a real leachate solution that contained many impurities. Specific energy consumption is observably inversely proportional to the current efficiency. This is related to the excessive usage of the current at 200 mA, as was observed by the formation of hydrogen gas at the anode. Selection of the applied current density denotes a balance between deposition kinetics and the power consumption. Conclusively, we found that 100 mA is the most efficient current to selectively deposit copper on the cathode in our experimental conditions.

7.4.3 **The final products**

Higher removal efficiency and selectivity was achieved in our experiments, owing to highly acidic conditions where the solubility difference of covellite enabled selective precipitation. The sulfidic precipitates, purity of 33% wt copper, were observed suitable to be used in the preceding smelting or roasting processes. The precipitates were greater in purity than the findings of Chen et al. (2014) who used a fractional precipitation strategy to separate metals from a metal-rich leachate solution. The leachate was turbid and light brown before solid/liquid separation, due to the formation of very fine CuS particles. Covellite precipitates were aggregated, amorphous, and of various particles sizes, as could be observed in Figure 7-7a. These finding and the micrographs were similar to other authors' findings who used synthetic and real solutions (Sahinkaya et al., 2009; Villa-Gomez et al., 2014). A large number of miniscule sulfidic precipitates are formed during the process, particle size as small as 100 nm as a result of low solubility and highly charged surfaces (Chung et al., 2015). From a practical point of view, solid/liquid separation of the precipitates could be problematic as the particles showed various particle sizes, including very small ones.

This leads to significant practical challenges with respect to solid–liquid separation and subsequent recovery of the precipitate. In our tests, centrifugation was used to gather the precipitates of various sizes however this is not possible on a larger scale. It should be noted that the practical efficiency might be lower, due to reasons stated above, despite the high metal removal achieved in tests. This problem might be remediated by the usage of a different aqueous sulfide source. As shown by Chung et al. (2015), usage of $Na_2S \cdot 5H_2O$ showed reduced surface charge, thus a larger precipitate size.

Zeta potential tend suggests that the high supersaturation resulted in the rapid nucleation of a large number of small particles, followed by the adsorption of excess sulfide onto the particle surface, imparting the negative charge. The surface charge suppressed aggregation of the primary particles, resulting in a substantially reduced mean particle size. The small, highly charged particles were stabilized in suspension, which resulted in rather poor settling characteristics.

The cathodic depositions were suitable to be used in the subsequent purification step. Regarding the deposit quality, the copper electrodeposit obtained was high grade (64.5%, Figure 7-7b). Scanning electron microscopy demonstrated a thin, compact and uneven deposit. When carbon felt is used as cathodes the formation of nuclei on the carbon surface and a change in the morphology of the deposit occurs (Bolzán, 2013). Also, in practice oxidation of the copper deposit with atmospheric oxygen constitutes a main challenge in future application of the technology at a higher scale. Cathode design can be further optimized so as to increase efficiency, possibly by increasing the fibre width and/or adjust the anode distance to influence the metal film formation.

On the other hand, there exists a number of practical of challenges at the precipitation process, particularly where the recovery of valuable metal is required. On the other hand, selectivity of copper on the cathodic depositions was good, with little impurities of iron and other metals.

Competing ferric iron cations is a potential reason for impurities on the cathodic depositions (Das and Gopala Krishna, 1996). Deposition of copper on the graphite felt was on the individual fibers of the material on thin layers. This mediated the oxidation of the deposited metal with atmospheric oxygen even though the specimen was carefully stored in nitrogen atmosphere. This may be overcome by increasing the copper deposited on the electrodes. Another strategy is the usage hot pressing vacuum so as to anchor the deposited metal onto the carbon cathode (Wan et al., 1997). Nevertheless, the grade of the copper deposition was high enough to be used in a subsequent electropurification step.

7.5 Conclusions

This work compared two aqueous metal recovery techniques with a pragmatic approach.

- We have shown the selective recovery of copper from a metal-rich leachate solution using two different techniques. The experimental results showed the good selectivity of the both methods.

- At sulfidic precipitation, copper is recovered stoichometrically, at a Cu:S ratio of 1:1. The precipitates showed agglomeration with no visible crystal formations.

- At electrowinning, copper is recovered from the solution by applying the appropriate voltage at a level that will allow solely the deposition of this metal. Due to speciation of iron at as ferrous iron in the leachate solution, selective deposition of copper was achieved.

- Electrochemical deposition is observed to be superior due its technical feasibility, ease of practical application, environmental friendliness and integrability with the preceding leaching process in and integrated recovery route of copper from waste material.

Chapter 8.

Techno-economic assessment and environmental sustainability analysis of a newly developed metal recovery technology

Abstract

Techno-economic assessment and environmental sustainability analysis carried out early in research, development and innovation (RD&I) projects may lead emerging technologies toward a better environmental profile. This chapter describes a case study for anticipatory techno-economic and environmental sustainability analysis that incorporates technology forecasting, cost-benefit analysis, and comparative impact assessment applied to a novel metal recovery technology from electronic waste material. The main objective is to compare, from a techno-economic and environmental point of view, different alternatives to metal recovery from electronic waste. Three case studies for chemical, biological and hybrid treatment of electronic waste for metals recovery are drafted and their environmental sustainability was assessed. The initial capital investment and operational costs for future resource recovery plant from early stage have been estimated and a cost-benefit analysis has been carried out. The average value of metal recovery of 1 kg of discarded PCB was 11.83 EUR. Taking into account the revenues and the total capital investment costs, 5.1, 2.4 and 4.3 years of return of interest were calculated for the biological, chemical, and hybrid process alternatives, respectively. In general, chemical technologies performed better than biological technologies, in terms of metal extraction and recovery efficiency. Environmental sustainability assessment revealed that biological treatment of base metals in acidophilic consortium is more advantageous than chemical leaching, However, in case of gold leaching, chemical leaching was less impactful than biological leaching. Thus, a hybrid approach is recommended in order to find a balance between high metal recovery yield and environmental sustainability. This case study illustrates the potential for ex ante LCA to prioritize research questions and help to guide environmentally responsible innovation actions taken in development of an emerging metal recovery technology.

8.1 Introduction

Resource scarcity, secure supply of raw materials and their adverse environmental, economic and societal impacts are alarming issues of the current century (Liu et al., 2015). The demand for raw materials, usage of natural resources and the increasing pressure on natural habitats urge the assessment of the impact of human activities (Dewulf et al., 2015). This particularly applies to resource- and energy-intensive products and processes, as sustainable resource management and energy supply are of high importance on the environmental and social impact agendas. Researchers look for new ways to incorporate sustainability issues into the design of new processes; and ensure the sustainable procurement, use and disposal of materials (Hallstedt et al., 2013). Novel, eco-innovative, sustainable processes are not only necessary to meet the newly adopted stringent environmental regulations, but also to keep a competitive advantage in the global markets (Gargalo et al., 2016).

Including the sustainability dimension at the earliest stage of development is increasingly a demand in research projects. Environmental profile assessment of an innovative material or a novel process at an early stage of development is of fundamental importance in its future application (Tecchio et al., 2016). Life cycle assessment (LCA) is a widely used and standardized tool to evaluate the environmental impact of a product, a process or a service. It systematically compiles inventories of resources attributable to the supply of products and processes. Such tools enable the identification of the potential impact and environmental hotspots of future systems, and potentially help giving strategic decisions at an early stage (Chang et al., 2014). As an important environmental decision making tool, LCA assesses environmental profiles and prioritizes research directions, and mitigates technical risks for scale-up prior to significant investments (Mancini et al., 2015). Applying LCA early in a research, development and innovation (RD&I) process guides emerging technologies towards a better environmental profile (Wender et al., 2014).

Early research, testing and development of an emerging technology, when its characteristics and parameters are defined, is an essential phase for its techno-economic and sustainability assessment. At this stage a first attempt can be carried out to predict the environmental impact for certain components of processes at an early laboratorial research stage. LCA is very useful for providing information on resource-and sustainability-related issues, such as the criticality of raw materials used in the supply chains, decision between usage of certain elements, or impacts of certain elements of a process on the environment (Mancini et al., 2016). Decisions made at an early stage of a research at a low technology readiness level (TRL), when most of the final costs, functional requirements and environmental impacts are determined, will have a large impact later (Gavankar et al., 2015).

Nonetheless, there are a number of limitations to accurately predict the environmental profile of a future technology. Early stage research processes only have laboratory experimental data available. Namely, barriers related to data scarcity, rapid change of technology and prices, issues related to scale-up and sensitivity, functionality, isolation of environmental impacts from technical research inhibit the application of techno-economic and sustainability analysis (Biddy et al., 2016). This task becomes even more difficult when data for a process that will occur in the future is needed. Upscaling from small to large scale applications shows significant scale up effects, as upscaling might give biased results which do not represent the performance of a technology in a future industrial scale (Rosner and Wagner, 2012). LCA is inherently conjunctive and cumulative, which brings into question the validity of the linearly scaled up data from the early stage low TRL development process (Villares et al., 2016). This does not, however, necessarily mean the invalidity of the results from the exploratory studies. It is essential to interpret the LCA results as a comparative analysis of systems but not a final verdict. For early-stage development of conceptual process alternatives, a multi-level framework for environmental sustainability and techno-economic analysis through multi criteria risk assessment is valuable (Herva and Roca, 2013).

LCA is retrospective in its nature, almost exclusively applied to existing products and services. Compared to ex post applications, LCA of future scenarios, i.e. ex ante cases require equal or occasionally more elaborate data input, as more uncertainties are involved. A number of issues are to be taken into consideration to draw sound conclusions, namely related to defining the goal and scope of the LCA, the quality of the data, and the level of confidence in data interpretation, (Chang et al., 2014). Attempts have been made to apply the tool to emerging technologies where requisite information for modelling is limited. Such approaches typically involve scaling up the technology, using scenarios based on estimates or simulations, where the associated production maturity and efficiency brings often more environmental benignity mainly as a result of the scale up effect (Gavankar et al., 2015). Thus the outcome of LCAs on emerging technologies based on data from lab scale should be interpreted in conjunction with their TRL. An LCA on a lab scale technology is of limited use for comparative purposes.

In technology development projects, scale-up is a decisive and integral step. The inherent lack of data across the life cycle of emerging technologies contributes to high uncertainty. Addressing differences between laboratory systems and industrial processes are crucial to establishing data validity. At low TRL lab phase, volumes are typically lower, where efficiency gains have been integrated (Frischknecht et al., 2009). Scaling up can uncover unforeseen byproducts or amplify small variations to large data errors. In the laboratory, the various steps are run in batch setting and not necessarily connected to each other and the volumes are much smaller than typical commercial plants. To the best of our knowledge, there is no general procedure to scale up laboratory scale processes to an industrial production. According to Piccinno et al. (2016), the scale-up framework follows a five-step procedure. The starting point **(1)** is a lab protocol that is obtained from the lab experiments directly, validated by a publication or a patent document, leading to a **(2)** design of a plant flow diagram. In the next step, each individual process step **(3)** in this plant flow diagram is scaled up according to the procedure of this framework. The **(4)** linkage and consolidation of the in- and output data of all the involved process steps is then included. All the obtained results are used in the concluding step **(5)** to perform the LCA.

Predicting future technology changes, which is an important factor in LCA results, and their associated environmental profile is difficult. Providing some reasonable consideration and appraisal of environmental impacts as early as possible in technological development is necessary, given the high degree of influence in such outcomes of this phase. Alternative process configurations, new interpretation methods capable of reconciling trade-offs between impact categories or technology alternatives can be beneficial in comparative LCA of emerging technologies. LCA analyses lacking explicit interpretation of the degree of uncertainty and sensitivities are of limited value as robust evidence for comparative assertions or decision making (Guo and Murphy, 2012). Uncertainty combined with the sensitivity analysis can lead to an incremental increase in confidence in the LCA findings, in particular in ex ante situations. The goal and scope must clearly specify the interpretation of results and their usage in defining environmental bottlenecks to compare the newly developed process routes (Hetherington et al., 2014).

The overall objective of this work is to conduct an environmental sustainability assessment and techno-economic assessment of an emerging future technology for the recovery of metals from electronic waste. Waste electronic and electronic equipment (WEEE) are an increasing fraction of municipal waste and an alarming global environmental problem. Their improper management leads to many environmental and social problems, i.e. such as pollution, informal recycling under substandard conditions, illegal transboundary movement of waste and loss of valuable resources (Tsydenova and Bengtsson, 2011). Recovery of metals from WEEE is an obvious option to meet the resource demands (Gu et al., 2016). A mobile phone includes more than 60 elements embedded within the device (Bloodworth, 2014), of which a large fraction is not properly recycled (Ueberschaar and Rotter, 2015). Various recovery routes have been considered to recovery metals from WEEE, including hydrometallurgical, biotechnological, pyrometallurgical approaches, and a number of hybrid approaches (Hadi et al., 2015; Hennebel et al., 2015; Zhang and Xu, 2016).

In Chapter 5 and 6, two proof-of-concepts of technological procesess to recover metals from waste material at lab scale was given. Namely, we developed a biological (Chapter 5) and a chemical (Chapter 6) process to selectively recover metals from discarded electronic devices. In this work, we aim to assess the sustainability profile and evaluate the techno-economic performance of a future technology. In this direction we developed three scenarios, namely a biological route, a chemical route, and a hybrid route, so as to comparatively analyse their profile from a life cycle point of view. Moreover, specific objectives are **(1)** to identify the key elements and bottlenecks for successful implementation, **(2)** give a comparative analysis of the selected scenarios, and **(3)** apply LCA as a tool for a strategic environmental sustainability analysis in technology development at an early stage.

8.2 Methodology

8.2.1 Simulation of an ex ante metal recovery process

This study evaluates the techno-economic profile and environmental sustainability of a future metal extraction and recovery process from WEEE, specifically printed circuit boards. This includes pretreatment, metal extraction, and a subsequent metal recovery steps, in a two-step approach, for copper (Cu) and Au (gold). Three alternative processes were designed, namely **(1)** a biological route, **(2)** a chemical route, and **(3)** a hybrid route. In each scenario, the study was conducted for hypothetical cases, with fixed amount of WEEE to be processed through slightly different extraction and recovery routes. The ex-ante process simulation was carried out by developing a flowsheet of a WEEE metal recovery plant in three alternative routes.

A flow diagram of the proposed processes, demonstrating the pretreatment, extraction and recovery stages in given in Figure 8-1. A throughput rate of 10 t/h of discarded PCB material was modelled for each process on which the techno-economic assessment was based. The distribution of various elements (chemical leaching, bioleaching, electrowinning) in the proposed routes was modelled and predicted from the equilibrium calculations and from primary laboratory data (Chapters 5, 6 and 7). In the actual resource recovery process, operational parameters such as temperature, concentrations, inflow rate, flow patterns, and tank geometry can significantly affect the distributions and final metal recovery from WEEE. Thus, it is worthwhile to note that the metal flows were predicted from stoichiometric calculations and could differ in the actual process. Nevertheless, the values presented in this chapter provide the limiting conditions which are based on a rigorous fundamental basis. The main goal of the study is to develop a framework for a future metal recovery plant and define the main elements of each alternative.

The feed is composed of discarded PCB from various electronic scrap. There is wide range of reported values for the metal concertations of discarded PCB (Chapter 4, Bigum et al., 2012; Fowler et al., 2015; Oguchi et al., 2012; Yamane et al., 2011). An average value was taken in order to simulate an actual full size PCB recycling plant receiving all types of WEEE. PCB are an integral type of every electronic device with varying metal concentrations, thus in practice much more variation might be expected. The composition of the throughput material is given in Table 8-1.

Table 8-1 Composition of the discarded PCB.

Metals	Cu	Au	Ni	Al	others	Total
Concentrations (mg/g PCB)	225	0.250	7.5	45	100	378
Value of the metals	2.15	11.35	0.1	0.05	0.1	11.86
Intrinsic value (%)	18.7	79.8	0.2	0.1	0.2	100

Copper (Cu) and gold (Au) are the most important metals to be recovered from discarded PCB, and they constitute 98.6% of the total value combined. Thus, the recovery processes are designed to efficiently recover these two metals in a two-step extracting and a subsequent selective recovery step, with the current level of information (Chapters 5 and 6). In the first step, Cu, along with other base metals is extracted in oxidative acidic medium, mediated either biologically or chemically. In the subsequent leaching step, copper-leached residues are subjected to Au leaching either in biologically mediated cyanide leaching, or chemically controlled ammoniacal thiosulfate medium. Both leaching steps were followed by a selective recovery step, namely electrowinning for the selective recovery of Cu and activated carbon adsorption for Au. The proposed routes are explained further in detail in the following sub-sections.

8.2.1.1 *The biological route*

In the biological route crushed PCB are fed to a bioleaching plant operated by acidophilic bacterial bioleaching consortium (Chapter 5). The particle size <500 μm was used in the process for two reasons: First is that after comminution most of the weight is accumulated in that particle size (Chapter 4) and the second is that the bacterial metal extraction is most effective in this particle size (Wang et al., 2009). The pretreatment (crushing) was carried with a rotary crusher, a total 7% loss was assumed (Chapter 4, Martins, 2016).

The tank is fed with Fe^{2+} and S^0 as an energy source for the acidophilic bacteria with externally added iron sulfate heptahydrate ($FeSO_4. 7H_2O$) and elemental sulfur (S^0), respectively. The bioleaching tank is sparged with air consisting of enriched oxygen (O_2) and carbon dioxide (CO_2) for two reasons: **(1)** facilitation of oxidative leaching of the metals **(2)** and to supply essential O_2 and CO_2 for the microbial requirements (Witne and Phillips, 2001; Ilyas and Lee, 2014b). It should be noted that the bioleaching bacteria are autotrophs, which means that they use atmospheric (non-fixed) CO_2 as their sole carbon source. The pulp density is selected as 10% (w/v), in accordance with the previously reported research work (Ilyas and Lee, 2014b; Mäkinen et al., 2015).

Following the bioleaching of base metals, the leachate solution is transferred to a subsequent electrowinning cell in which copper is recovered by applying electrical current (50 mA/cm^2, 6.42 kWh/gr Cu) to the solution (Chapter 7). The liquid is recycled back to the bioleaching solution to feed the bacteria with residual iron from the electrowinning step (Chapter 7). The product of this step is the recovered Cu in its elemental form. The residues from the first bioleaching step are sent to a secondary bioleaching step by cyanogenic bacteria with 5% pulp density (w/v) (Chapter 5). The tank is fed with glycine and standard growth medium for the production of cyanide (Chapter 5). Leached Au is recovered in an adsorption tank using activated carbon (15 g/L) (Navarro et al., 2006).

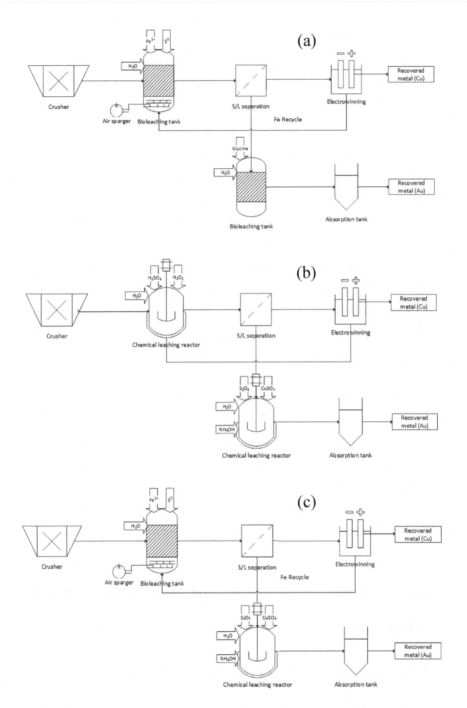

Figure 8-1: Schematic process design of (a) biological, (b) chemical, and (c) hybrid routes for future process routes of metals recovery from electronic waste.

8.2.1.2 *The chemical route*

In the chemical route, Cu is leached by a mixture of sulfuric acid (H_2SO_4) and hydrogen peroxide (H_2O_2). Cu reacts very poorly with solely H_2SO_4 and requires an oxidizing agent for the efficient extraction of this metal from electronic waste (Yang et al., 2011). Thus, the chemical leaching tank required dosing of optimized amounts of H_2SO_4 (3.93 M) and H_2O_2 (3.92 M) (Chapter 6). The leachate solution from the oxidative leaching is then subjected to a selective electrowinning step for the recovery of Cu as similar to biological route. The copper-leached residues, on the other hand, are leached by thiosulfate ($S_2O_3^{2-}$) in an ammoniacal (NH_4^+) solution, catalyzed by cupric ions (Cu^{2+}), for Au extraction. The operational process parameters are selected according to the optimized concentration of $S_2O_3^{2-}$ (0.3 M), $CuSO_4$ (0.038 M), and NH_4OH (0.38 M) (Chapter 6). In the subsequent adsorption step on activated carbon, a higher concentration (60 g/L) was used compared to the biological route, due to higher concentration of Au in the ammoniacal solution. This process includes techniques for eluting the gold from the resin and restoring the resin to recycle back to the adsorption circuit.

8.2.1.3 *The hybrid route*

In the hybrid route, the approach is to combine the biological leaching of Cu by iron-and sulfur oxidizer acidophilic bacteria and chemical leaching of Au in an ammoniacal thiosulfate (NH_4OH-$S_2O_3^{2-}$-$CuSO_4$) solution. The bioleaching tank is fed with optimized concentrations of iron sulfate heptahydrate and elemental sulfur. Subsequently Cu and Au are recovered by low current electrowinning and activated carbon absorption (60 g/L), respectively, similar to the other alternative routes. As similar in the bioleaching route, the leachate solution after electrowinning will be recirculated back to the bioleaching step to provide the bacteria with an additional ferrous iron (energy source).

8.3 Sustainability assessment of the newly developed technology

8.3.1 Description of the life cycle assessment system

The goal of the ex-ante life cycle assessment (LCA) is to carry out a comparative analysis of the potential environmental impacts of the three scenarios for metal recovery from WEEE. The system was designed and modelled to expand the focus on a pretreatment (crushing) and a two-step extraction and a recovery step of copper (Cu) and gold (Au) in three different scenarios, namely, chemical, biological and hybrid routes (See 8.2.1). The side objectives are: **(1)** the appropriate selection of the environmental indicators, **(2)** the inherent data gaps and strategies to address this issue, **(3)** the identification of hotspots in each alternative at early stage of development and **(4)** how life cycle impact assessment results can influence the technology development research itself. The outcome of the ex-ante LCA establishes the environmental hotspots and potentially reorients the research direction.

An attributional LCA of a product system, made up of the unit processes was performed in order to map the cumulative environmental impact profile of each process. The primary laboratory data was scaled to an industrial size applying a plausible scenario. An environmental impact performance comparison of the three scenarios was carried out applying the four phases of the life cycle assessment framework, namely **(1)** Goal and scope definition, **(2)** Life cycle inventory (LCI) analysis, **(3)** Life cycle impact assessment (LCIA), and **(4)** Interpretation (Guinée et al., 2002; ISO 14044, 2006). LCA was a prospective attributional forward looking and descriptive ex ante LCA. A scenario was applied to define a relevant future state of metals recovery from WEEE. The system boundaries excluded the unit processes but not affected by the reference flows of the metal recovery processes. To the extent possible, the foreground system was modelled with specific data. The level of sophistication was as comprehensive as possible.

A cradle-to-grave approach is adopted for the LCA case-study of the three processes. The functional is the extraction and recovery of elemental Cu and Au from WEEE. The functional unit is selected as a physical unit and 1 kg of discarded PCB waste for recovery. The major components of the process are the pretreatment of the discarded material, and the extraction and the recovery of the metals in a two-step process. The crushing of the waste material, the chemical and biological metal extraction (leaching) in tanks, the subsequent recovery (electrowinning and adsorption), the waste water treatment, and the disposal of the final residues processes were included in the LCA system boundary. An illustrative schematic of the LCA flowchart is given in Figure 8-2.

8.3.2 Inventory analysis and data collection

The European Life Cycle Database (ELCD) v3.2.0 was used in the OpenLCA (v1.5.0) software for the LCA analyses. The site-specific inventory data is used where possible, data for the Netherlands have been used for the background processes, if available in the database. When unavailable either regional European data were used as a reasonable alternative. The product systems are from cradle to gate, in which the cradle is the end of life EEE. Extraction of resources and emissions from previous life cycle are not taken into consideration. Neither the manufacture nor the use phase of electrical and electronic equipment (EEE) are incorporated as the origin of the material flow. The pretreatment (crushing), metal extraction and metal recovery stages of the system derived from background information from primary data and ILCD data supplemented with assumptions and estimations. The latter came from the generic data relating to metal recovery processes. Completeness of the product systems had a priority over consistency as without the former the information obtained was not of value. The product systems are defined as completely as possible in order to simulate a realistic future metal recovery system.

8.3.3 Life cycle impact system, classification and characterization

A broad range of relevant categories and indicators are applied to cover ecosystems, human health, and resources depletion. Applied impact categories are given below in Table 8-2. The impact assessment method is the mid-point level of the cause effect chain on the environmental impacts. OpenLCA Life Cycle Impact Assessment (LCIA) methods (v1.5.5) were used for the analysis of the impacts of the selected indicators. Classification whereby the environmental impact from the inventory analysis results are assigned to the chosen impact categories is computed in the OpenLCA (v1.5.0) software. In the characterization step, environmental impact and characterization models are multiplied and aggregated and gave the category indicator results. The impact results were analyzed for the product systems built for the three recovery routes.

8.3.4 LCA data quality

The LCA inventory was developed by using site-specific primary data collected from industrial sources and primary data from laboratory experiments supplemented with secondary data from publicly available sources and the European Life Cycle Database (ELCD) (v3.2). The primary datasets were developed with the available primary data (Chapters 4, 5, 6 and 7) and from literature sources. First using primary data provided by laboratory experiments of metal recovery from printed circuit boards (PCB) was modelled, where available.

Table 8-2: Impact categories of life cycle assessment of metal recovery processes.

Impact category	Impact group	Category indicator	Characterization factor	Unit
Depletion of abiotic resources	Resources	Ultimate reserve related to annual use	Abiotic depletion potential	Kg-antimony-eq
Depletion of water resources	Resources	Ultimate reserve related to annual use	Abiotic depletion potential	m^3
Eutrophication potential	Ecosystem health	N and P emissions to biomass	Eutrophication potential EP	Kg-PO_4-eq
Acidification potential	Ecosystem health	Deposition/acidification critical load	Acidification potential: AP	Kg-SO_2-eq
Climate change	Ecosystem health	Infrared radiative forcing	Global warming potential GWP 100a	Kg-CO_2-eq
Photochemical oxidation (smog)	Human health	Tropospheric ozone formation	Photochemical oxidant creation potential	Kg-ethylene-eq
Stratospheric ozone depletion	Ecosystem health	Stratospheric ozone breakdown	Ozone depletion potential	Kg-CFC-11-eq
Ecotoxicity - terrestrial	Ecosystem health	Predicted environmental concentration	Terrestrial ecotoxicity potential: TAETP infinite	Kg-1,4-DCB-eq
Ecotoxicity – marine aquatic	Ecosystem health	Predicted environmental concentration	Marine Ecotoxicity potential: MAETP infinite	Kg-1,4-DCB-eq
Ecotoxicity – freshwater aquatic	Ecosystem health	Predicted environmental concentration	Marine freshwater potential: FAETP infinite	Kg-1,4-DCB-Eq

Available experimental data included **(1)** crushing of the discarded board material, **(2)** bioleaching of Cu with *Acidithiobacillus ferrivorans* and *Acidithiobacillus thiooxidans*, **(3)** bioleaching of Au with *Pseudomonas putida*, **(4)** chemical leaching of Cu with H_2SO_4 and H_2O_2, **(5)** chemical leaching of Au with $S_2O_3^{2-}$, $CuSO_4$ and NH_4OH, **(6)** electrowinning of Cu. The data gaps were filled using information from the literature., e.g. adsorption data for Au recovery with activated carbon was taken from Syed (2012), and air sparger consumption data for was taken from Witne and Phillips (2001).

8.3.5 Sensitivity analysis and uncertainty

LCA studies should communicate the reliability of their results in terms of uncertainty based on an assessment of the data quality of the information used (Bieda, 2014; Weidema, 2000). This means that a single deterministic LCA result might not be enough for a clear understanding of the environmental performance of a system. Some inconsistencies inherently occur in the build-up of the system. The scenario sensitivity analysis method suggested by Björklund (2002) was applied in this work. This method involves calculating different scenarios, to analyze the influence of input parameters on either LCIA output results or rankings. At LCI level, the uncertainty introduced into the inventory due to the cumulative effects of input uncertainty and variability of inventory data was quantified by using either statistical methods or the expert judgement-based approach. Sensitivity analysis on the allocation method was not undertaken in the current study, but can be explored in further research.

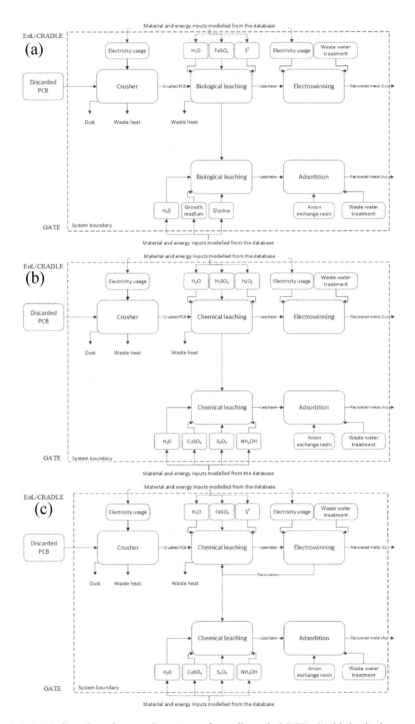

Figure 8-2: LCA flowchart for metal recovery from discarded PCB: (a) biological route (b) chemical route (c) hybrid route.

8.4 Techno-economic assessment of the processes

A comparative cost benefit analysis of the proposed processes is essential in order to establish a techno-economic assessment. The techno-economic analysis of the proposed process depends on the variable operational costs, fixed purchased equipment costs and based on the plant design. The main relevant economic parameters for the plant design are given in Table 8-3. The plant design assumptions were based on a realistic scenario, benchmarked by existing plants from waste collection and minerals processing sectors. The plant was designed to operate 300 days/year on a 20-hour based daily shifts, with a daily waste processing capacity of 10,000 tons and an economic life span of 20 years. The revenues are estimated based on the recovered metals taking into account the process metals recovery efficiencies and overall plant efficiency. The price volatility was not characterized. The economic values of the metals were collected from the from London Metal Exchange on 19 August 2016 (LME, 2016) on a year-basis period, between 1 August 2015 and 1 August 2016.

A conceptual economic evaluation of the processes alternatives was conducted based on cost information of available metal production plants. The techno-economic assessment was performed through a procedure comprising of a process simulation followed by an economic cost benefit analysis of the variable operational and fixed cost analysis of each step. The revenues were calculated according to the metal concertation of the boards (Table 8-1) and the potential copper (Cu) and gold (Au) recovery yield taking into account the process efficiency and overall plant efficiencies of each alternative process. The sensitivity analysis of the scale-up procedure was not elaborated. The primary data were extrapolated to full-scale plants, and the total efficiencies were calculated taking into account the overall plant efficiency.

Total capital investment (TCI) includes all costs required to purchase equipment needed for the control system (purchased equipment costs), the costs of labor and materials for installing that equipment (direct installation costs), costs for site preparation and buildings, and certain other costs (indirect installation costs). TCI also includes costs for land, working capital, and off-site facilities. The overall variable and fixed operating costs, as well as the capital costs, were estimated for the three integrated process alternatives and their respective calculated flow rates. The variable operational costs are given in Table 8-4. The costs are calculated according to the monetary cost of the reagent, service or infrastructure. The prices of the chemicals were taken from an international internet marketplace for chemicals (alibaba.com, Cucchiella et al., 2015; Ghodrat et al., 2016). Electricity prices for the industry in the Netherlands are taken from the

European statistical office (Eurostat, 2016). Energy consumption for the mechanical crusher is taken from Martins, 2016; and Villares et al., (2016).

Table 8-3: Economic analysis for plant design.

Description	Units	Value
Annual working hours	h/year	8320
Plant available time	h/year	7,200
Plant capacity	ton/d	10,000
Overall plant efficiency	%	95
Plant life	years	20

The capital and operating costs in the order of magnitude were estimated using the sixth tenth rule: The calculation equation is given in Equation 8.1.

$$C_f = C_1 \times (\frac{S_1}{S_2})^{0.6} \qquad (8\text{-}1)$$

Where C_1 is the capital cost of the plant with capacity S_1 and C_f is the capital cost of the proposed plant with capacity S_2. Where there was no existing plant, the capital cost was estimated based on factors of the purchased equipment cost available in traditional design textbooks and technical literature. All inflationary adjustments were done using the 2015 Chemical Engineering Plant Cost Index (CEPCI) as well as those available in the literature.

Fixed capital costs were calculated based on the operational capacity, the process design, and the cost estimates for the major processing equipment and are given in (Table 8-5). The capital costs of the pretreatment (crusher) was calculated on the basis of data available for a medium-sized processing plan in Europe (Neto et al., 2016). The throughput (10 t/day) corresponded to 162 man-hours/day of manual dismantling, based on the manual dismantling rate calculated by Fan et al. (2013). For the remaining process section, the bioleaching tank was estimated based on a literature report scaled to the production rate by the verified stoichiometry (Brierley and Brierley, 2013). Similarly, the chemical leaching tank estimation was done based on earlier work published by Rocchetti et al. (2013). For electrowinning, the primary experimental data from the setup was straightforwardly extrapolated (Chapter 7), leading to a design with cathode Cu production capacity of 197.6, 195 and 197.6 kg Cu/h with a consumption of 5.24 kWh/kg Cu.

Table 8-4: Variable operational costs of the three alternative routes metal recovery processes.

Reagent costs				
Description	**Usage (kg/kg PCB)**	**Price (EUR/kg)**	**Cost (EUR/kg PCB)**	**Used process route(s)**
Iron sulfate heptahydrate ($FeSO_4 \cdot 7H_2O$)	0.087	0.07	0.0056	1, 3[§§]
Elemental sulfur (S^0)	0.01	0.14	0.0014	1, 3
Glycine (NH_2CH_2COOH)	0.01	1.07	0.0107	1
Sulfuric acid (H_2SO_4) (70%, v/v)	0.082	0.18	0.0144	2
Hydrogen peroxide (H_2O_2) (35%, v/v)	0.064	0.36	0.0216	2
Copper sulfate pentahydrate ($CuSO_4 \cdot 5H_2O$)	0.017	1.48	0.0252	2, 3
Sodium thiosulfate pentahydrate ($Na_2S_2O_3 \cdot 5H_2O$)	0.086	0.31	0.0267	2, 3
Ammonium hydroxide (NH_4OH)	0.019	0.22	0.0042	2, 3
Activated carbon (C)	0.012	1.43	0.0172	1, 2, 3
Electricity costs				
Description	**Usage (kWh/kg PCB)**	**EUR/kWh**	**Cost (EUR/kg PCB)**	**Used process route(s)**
Electricity (Crushing)	0.038	0.084	0.0319	1, 2, 3
Electricity (Electrowinning)	1.24	0.084	0.104	1, 2, 3
Electricity (Aeration)	0.02	0.084	0.017	1, 3

[§§] 1: Biological route, 2: Chemical route, 3: Hybrid route

The activated carbon adsorption tank was modelled based on an earlier work carried out by Navarro et al. (2006), taking into account the Au extraction rate of the corresponding leaching processes.

Total capital investment costs (TCIC) are given in Table 8-6. They included the total direct installation costs (TDIC) and total indirect costs (TIIC) plus fixed capital investment costs (FCI). Total investment costs are estimated based on the assumption given in Chemical Engineering Plant Cost Index (CEPCI), and by scale-up procedures and factors given by Picciono et al. (2016) and Biddy et al. (2016). TDIC included infrastructure costs, foundations and supports, installation of the equipment, electrical work, office buildings, and piping, and calculated as a function of the TPEC. TIIC included engineering and design costs, construction costs, field expenses, contractor fees, start-up and performance test costs and contingencies. Contingencies is a group of costs that covers unforeseen costs such as possible redesign and revision of project components, modification of the equipment, escalation increases in cost of equipment, increases in field labor costs, and delays encountered in the installation. Contingencies must not be mistaken for uncertainties and retrofit factor costs.

8.5 Results

8.5.1 Economic evaluation

A mass balance flowsheet of copper (Cu) and gold (Au) extraction and recovery from discarded PCB is given in Figure 8-3. The potentially recovered metals are calculated to be 197.6, 195, 197.6 g/kg and 0.102, 0.226 and 0.224 g/kg Au, respectively, taking into account the process efficiencies for the biological, chemical and hybrid alternatives and overall plant efficiency. Chemical leaching had the higher gold recovery efficiency, owing to a relatively higher process efficiency. The gold recovery of the hybrid process was slightly lower than the chemical process route (0.224 to 0.226 g Au/kg PCB), whereby its Cu recovery was identical to the bioleaching route (197.6 gr Cu/kg PCB). The potential revenues were calculated to be 5.03, 9.64 and, 9.26 EUR/kg PCB, respectively, for the biological, chemical and hybrid alternatives.

Table 8-5: Total purchased equipment costs (TPEC) estimations.

Unit	Capacity	Units	Costs (EUR/unit)	Total cost (EUR)
Biological route				
Crushing	500 kg/hr	5	20,000	100,000
Bioleaching tanks	5,000 m³	20	7,500	150,000
Electrowinning cells	10,000 m³	8	25,000	200,000
Adsorption tanks	5,000 m³	15	5,000	75,000
Total purchased equipment costs (TPEC)				**525,000**
Chemical route				
Crushing	500 kg/h	5	20,000	100,000
Chemical leaching tanks	13,000 m³	8	7,500	60,000
Electrowinning cells	10,000 m³	6	25,000	150,000
Adsorption tanks	5,000 m³	12	5,000	60,000
Total purchased equipment costs (TPEC)				**370,000**
Hybrid route				
Crushing	500 kg/hr	5	20,000	100,000
Bioleaching tanks	5,000 m³	20	7,500	150,000
Chemical leaching tanks	13,000 m³	8	7,500	60,000
Electrowinning cells	10,000 m³	8	25,000	200,000
Adsorption tanks	5,000 m³	15	5,000	75,000
Total purchased equipment costs (TPEC)				**585,000**

Table 8-6: Total capital investment costs (TCIC) calculations.

Cost description	Calculation method	Biological route (EUR)	Chemical route (EUR)	Hybrid route (EUR)
Total purchased equipment costs (TPEC)		**525,000**	**370,000**	**585,000**
Installation factor (IF)		2.0	2.0	2.0
Total direct installation costs (TDIC)	TPEC × IF	**1,050,000**	**740,000**	**1,170,000**
Engineering and design	0.32 × TPEC	168,000	118,400	187,200
Construction expenses	0.34 × TPEC	178,500	125,800	198,900
Contractor fee and legal expenses	0.23 × TPEC	120,750	85,100	134,550
Contingencies	0.20 × TPEC	105,000	74,000	117,000
Total indirect investment costs (TIIC)		**572,250**	**403,300**	**637,650**
Total depreciation capital (TDep)	TDep = TDIC + TICC	**1,662,250**	**1,143,300**	**1,807,650**
Labor capital	36,000 man-hours	360,000	360,000	360,000
Royalties	2.8% of TDep	46,543	32,012	50,614
Land (real estate)	2.8% of TDep	46,543	32,012	50,614
Fixed capital investment costs (FCI)		**453,086**	**424,024**	**461,228**
Total capital investment costs (TCIC)	TPEC + TDIC + TIC + FCI	**2,745,336**	**2,677,324**	**4,661,528**

The revenues are calculated as per the recovered Cu and Au, taking into account the process efficiencies for each alternative. The total costs of the process were calculated summing the total capital investment costs (TCIC) and total operational costs (TOC). The total costs are given in Table 8-7.

Table 8-7: Total costs of the three metal recovery routes.

Process route	Total operational costs (EUR/kg PCB)	Total capital investment costs (TCIC) (EUR/kg PCB)	Total costs (EUR/kg PCB)
Biological route	0.159	0.457	0.616
Chemical route	0.224	0.446	0.670
Hybrid route	0.232	0.776	1.008

Table 8-8 shows the cost benefit calculations of the metal recovery plants, taking into account the value of the discarded PCB, and the potential revenues based on the recovery of Cu and Au from the discarded material. The total variable costs of the process alternatives are 0.616, 0.670 and 1.008 EUR/kg PCB, respectively for the biological, chemical and hybrid process routes. An annual revenue of 1,324,000, 2,691,000 and 2,475,000 EUR, respectively, for the biological, chemical and hybrid process routes was calculated. Taking into account the revenues and the total capital investment, the a 5.1, 2.4 and 4.3 years of return of interest was calculated for the biological, chemical, and hybrid process alternatives.

Table 8-8: Cost-benefit analysis of the metal recovery processes.

Value (EUR/kg)	Process route	Plant efficiency (%)	Recovery efficiency (%) Cu	Recovery efficiency (%) Au	Potential revenue (EUR/kg PCB)	Total costs (EUR/kg PCB)	Net revenues (EUR/kg PCB)
	Biological		87.5	34.0	5.03	0.616	4.414
11.83	Chemical	95	86.6	75.3	9.64	0.670	8.97
	Hybrid		87.8	74.6	9.26	1.008	8.250

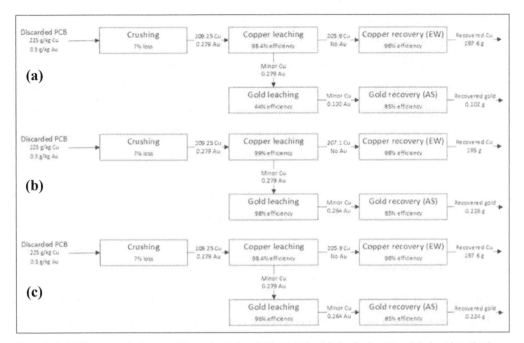

Figure 8-3: The mass balance of copper and gold in (a) the biological route, (b) the chemical route, and (c) the hybrid route.

8.5.2 Environmental sustainability

8.5.2.1 *Life cycle impact assessment*

The comparative results of the life cycle impact assessments (LCIA) of the three processes are given in Table 8-9. The analysis was carried out for 1 kg of waste PCB treated for metal recovery. The main goal was to obtain comparable results of the alternative process routes and draw conclusions based on the individual process contributions on the selected indicators. The results are given according to individual selected impact categories and the type of allocation (mass). Only the environmental impact of the recovery of metals from high-grade PCB was evaluated, and the removal and subsequent treatment of hazardous components and plastic was excluded, except for the treatment of final metals-extracted residues and wastewater treatment of the leachate solutions. The study does, therefore, not represent the overall environmental cost of treating 1 kg of high-grade PCB, but represents a comparative environmental impact assessment of Cu and Au recovery in various alternative process routes.

The overall process contribution to climate change was 8.46. 14.6 and 11.6 kg CO_2-eq/kg PCB processed for the three alternative process routes namely, the biological, the chemical and the hybrid routes. The main causes were the mixed electricity usage (in the Netherlands), which came largely from the combustion of coal (results not shown). Moreover, the assessment showed significant environmental impacts on emissions to the water resource, e.g. acidification potential and resources depletion. Overall, the chemical waste processing route had a higher impact over the biological in categories in all categories except for freshwater eutrophication, land use, ozone depletion, and particulate matter/respiratory inorganics. The impact of the chemical route om water resources depletion was considerably higher (%131) than that of the biological route The impact of the hybrid route was higher than the biological route, but lower than the chemical route in all impact categories.

8.5.2.2 *Process contributions*

The results of the individual process contributions of three alternatives are given in . Electrowinning had a higher impact on climate change and greenhouse gas (GHG) emissions related impacts, such as climate changes, ionizing radiation, photochemical ozone formation in all three alternatives. Also, the contribution of electrowinning on acidification, freshwater eutrophication and land use was impactful, contributing to the 2nd highest process to these impact categories in all three process alternatives. In water-resources impact factors, e.g. freshwater eutrophication, water resources depletion, and acidification, metal extraction processes, i.e. bioleaching, chemical leaching, were the main contributors. The impact of chemical copper leaching on freshwater eutrophication was much higher (46.4% to 24.2%) than the biological leaching of copper. The attributional impact of chemical and biological leaching of copper was comparable when water resources depletion was concerned. Similarly, the impact of chemical gold leaching on the water resources depletion was lower than that of biological leaching, owing to the relatively low water consumption of the chemical process alternative.

Table 8-9: Life cycle impact assessment results of the three routes.

Indicator	Group	Unit	Biological route	Chemical route	Hybrid route
Freshwater eutrophication	Ecosystem health	kg P-Eq	0.216	0.026	0.121
Land use	Ecosystem health	kg SOC	1.98	1.4	1.69
Ionizing radiation, ecosystems	Ecosystem health	CTUe	2.10E-03	2.40E-03	2.25E-03
Climate change, GWP 100a	Ecosystem health	kg CO_2-Eq	8.4	14.6	11.6
Ozone depletion	Ecosystem health	kg CFC-11-Eq	8.60E-03	8.40E-03	8.50E-03
Acidification potential	Ecosystem health	kg SO_2-Eq	0.160	0.744	0.424
Human toxicity, carcinogenic	Human health	CTU-h	12.4	28	20.2
Particulate matter/respiratory inorganics	Human health	kg PM 2.5-Eq	0.41	0.3	0.36
Photochemical ozone formation	Human health	kg ethylene-Eq	2.40E-03	3.60E-03	3.00E-03
Resource depletion, mineral, fossils and renewables	Resources	kg Sb-Eq	0.64	0.842	0.541
Resource depletion, water	Resources	m^3	11.4	26.4	14.4

Crushing had the largest impact in all three processes on the particulate matter/respiratory inorganics indicator, mainly due to the dust emissions to the air. This unit is common in all the process alternatives and its impact on the categories were similar for all the process alternatives, sharing a comparable fraction of the impact assessment (). The fraction of crushing in the selected categories was mostly dependent on the other process units. When the impact of other units, e.g. bioleaching or chemical leaching increased, the relative impact of this process of crushing was lower. Similarly, despite the fact that the impact of crushing was identical in three process alternative, its effect on the on climate change and ionizing radiation was larger for the biological route, due to lower individual effect of other units on climate change in this process alternative. The Au absorption had a significant impact on photochemical radiation, ionizing radiation, and freshwater eutrophication. Its relative fraction on these categories were comparable, however slightly smaller for photochemical ozone formation and ionizing radiation and higher for freshwater eutrophication compared to the biological and hybrid process alternatives, mainly due to varying impacts of other processes. It is extremely important to include scenario analyses in the scale-ups to understand the implications of assumptions and manage the uncertainty. Beyond the results of the environmental impacts, this scale-up framework offers further advantages: simple estimations of the variable production costs become possible. The results of the scale-up depend on the knowledge and the data quality that is applied, scaling the process approximately using the same steps as in the laboratory scale. Scenario analyses are therefore an important and decisive aspect for the robustness and credibility of a scale-up study. The limitations or open points of the approach mainly regard its limited applicability and the data quality. A scale-up based on the laboratory experiments might be useful in justifying the research of a new material or process by showing its potential environmental performance. In practice, the sensitivity of the results generally varies at the same time, so different scenarios in which the most significant parameters analysis are varied simultaneously are considered to determine the impact on the annual costs, revenues and net benefits from the metal recovery from WEEE.

8.5.3 Techno-economic analysis and feasibility of the processes

Technology development processes are usually driven by an economic motivation and the feasibility of a proposed action must be proven before taking any scale up operations. Construction of an integrated plant for recycling metals is a very large investment, and will only be undertaken where it has 'economic feasibility', where the desired return on investment is reliable and not subject to strong risk (Ghodrat et al., 2016). In the event of a trade-off between a cost efficient and an environmentally friendly process, the former will presumably be chosen. In the best case scenarios, a newly proposed technology is both cost-effective and environmentally friendly, proving superior to the best available technologies (BAT). An opportunity is thus opened to compare a new production process to competing materials that are already on the market and produced at an industrial scale. It helps to assess the process itself by high-lighting hotspots and bottlenecks with high contribution to the variable and fixed operational costs. Further application and development will help in optimizing the results and the procedure of this framework and an expansion to more processes (e.g. continuous reactions, inclusion of gaseous and solid state reactions, cooling etc.) is desirable.

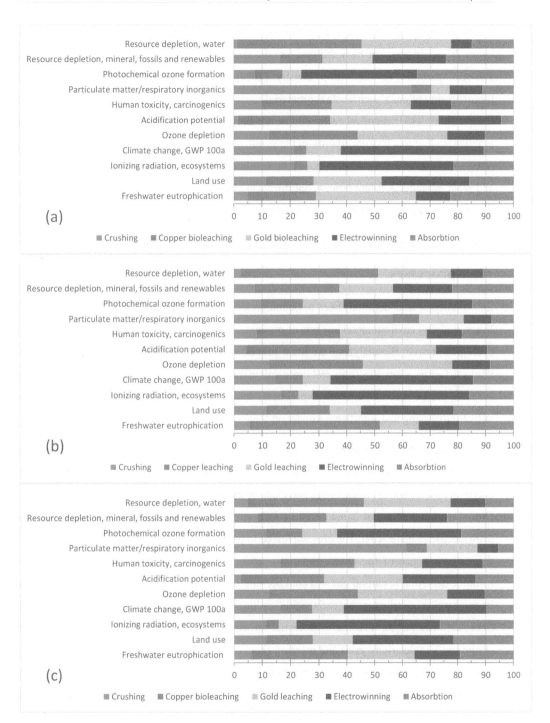

Figure 8-4: Process contributions of life cycle impact assessment (LCIA) of three alternative process routes, namely (a) the biological route (b) the chemical route, and (c) the hybrid route.

It is of no surprise that the chemical technologies perform efficient-wise and thus economically better. It should however be noted that that the simulation of biological in a highly optimistic scenario, assuming a high pulp density (10%) in light of the recent development in bioprocessing of metal-rich waste for material recovery (Makinen et al., 2015, Ilyas et al, 2014). On the other hand, chemical technologies are based on established process stoichmetries in practice leading to a less uncertainty in the simulation of scale up procedure. Pulp density is an integral parameter in process design, as it had collateral effects both on tank design, and volume and plant capacity. A future process optimization effort aiming to maximize process efficiency should prioritize the optimization of pulp density.

8.5.4 Environmental sustainability of the project metal recovery processes

Environmental considerations need to be integrated in decisions related to future technologies. When studying environmental impacts of products and services it is vital to study these in a life cycle perspective. By analyzing the main processes in metal recovery, we developed certain relevant factors - ranging from elaborate calculation procedures to qualitative guidance - to facilitate the LCA scale-up of lab experiments to an industrial scale. The results showed that WEEE processing had significant impact on the environment.

Potential eutrophication is largely derived from phosphate emissions to the ground mining from tailings from coal and ore mining. The coal is used as fuel in energy plants delivering electrical energy for the aeration (bioleaching), and electrowinning of the biological route product system. Global warming potential is increased by CO_2 emissions from the greenhouse gas emissions from electricity generation. The ores are linked to primary copper production also necessary for the very existence of the Dutch electricity grid. The main contributor to potential abiotic resource depletion is natural gas and coal burning for Dutch power plants. NO_2 emission from its combustion contribute to potential acidification. They are also responsible for photochemical oxidation. The largest addition to potential terrestrial ecotoxicity is caused by emissions of chromium VI (electricity grid) and mercury (coal combustion). The largest impacts on marine ecotoxicity also relate to upstream material inputs to the bioleaching system. The greatest contributing background unit processes are of global disposal of coal and lignite mining spills which produce emissions of nickel and beryllium. Potential freshwater ecotoxicity also shows up contributions from mining soil disposal. Potential human toxicity has large contributions from sulfuric acid production, thus the impact of chemical leaching was larger than that of biological leaching. Biological processing had also relative lower impacts to the freshwater resources depletion, owing to its lower water consumption, coupled with higher pulp density relative to chemical processing.

Taking into account the total impacts, the biological treatment option had lesser impacts on virtually all impact categories than the chemical treatment option. Similarly, the hybrid approach had lower impacts than the chemical approach on the impact categories related to the air emissions, water emissions and the natural resources depletion, e.g. as and climate change, freshwater eutrophication, acidification potential and resources depletion. The hybrid metal recovery approach, coupled the environmental benefits of copper sketching a profile to 'best of the two worlds' approach. The less-impactful biological treatment might be combined with high yield chemical treatment coupling the advantages of the two processing approaches. Despite the relatively higher total costs of the hybrid treatment approach, it might compensate with the highs with a better environmental profile.

8.6 Conclusions

The techno-economic assessment and environmental sustainability analysis of a metal recovery technology in development stage was studied. Three alternative process, namely biological, chemical and hybrid metals recovery routes were designed, simulated and analyzed, using primary data when available, otherwise assumed or taken from the literature. The results showed that the chemical metal technologies are more efficient than biological technologies in terms of final metal recovery. Moreover, thus the potential revenue from chemical technologies was higher than that of biological revenues. Techno-economic evaluation of the simulation of a full-scale plant showed that a 5.1, 2.4 and 4.3 years of return of interest for the biological, chemical, and hybrid process alternatives, respectively. The hybrid route had the highest capital investment and operating costs. In general, biological alternative had a lower environmental impact compared to chemical alternative in many impact categories. The high operating costs of the hybrid processing alternative can be combined with the less-impactful.

Chapter 9.

General discussion and conclusions

9.1 Introduction

Waste electronic and electrical equipment (WEEE) generation reached 41.8 M tons globally in 2014, of which 9.8 M, 7.2 M and 6.0 M belonged to EU-28, USA and China, respectively (Baldé et al., 2015; StEP, 2015). The generation will increase in the future, particularly in the developing economies (Chapter 2). It is a catastrophic environmental problem due to the hazards associated with its improper management including informal substandard treatment, illegal transboundary movement and trafficking, and disposure of the toxic substances into the environment. In addition to its hazards, WEEE is an important secondary source of materials, particularly metals. Many environmental concerns already arose associated to the limited space available for the final disposal of WEEE, the declining primary ore reserves and potential future deficit of technology metals, conditions in informal treatment establishments, and the need for control of environmental contamination.

Recycling and recovery of metals from WEEE is an obvious choice to meet the global demand for technology metals (Gu et al., 2016). 80% less resource consumption is achieved when desktop computers are recycled and the metals are recovered (Van Eygen et al., 2016). From a recycling point of view, WEEE is a mixture of polymetallic substances consisting of up to 60 metals, found mostly in their elemental form (dissimilar to the primary ores) along with plastics, silicates, mixed in a complex matrix (Chapter 2, Bloodworth, 2014). Particularly printed circuit boards (PCB) are an important source of copper (Cu) and gold (Au). The concentration of these metals in discarded PCB are many times higher than those of primary ores (Chapter 4, Akcil et al., 2015). This requires a novel approach to selectively and sustainably recover metals from this metal-rich anthropogenic secondary resource.

The complexity of the polymetallic WEEE, the concentration of the metals found in the waste material and their chemical properties distinguishes this material from primary ores. Technology selection for metal recovery from new WEEE streams must be done taking into account all pillars of sustainable development, including environmental, and social factors. The decision for technology selection should incorporate factors involving the characterization of the waste, available technologies and their recovery efficiencies, as well as their social implications and environmental impacts.

The current state-of-the-art WEEE recycling is limited to pyrometallurgical, and to a smaller extent hydrometallurgical approaches. Currently a number of smelting facilities, historically primary ore smelters operate in Europe that also process WEEE for metal recovery: Umicore in Belgium, Aurubis in Germany, Boliden in Sweden, and Glencore in Switzerland. The principle of pyrometallurgical metal recovery from WEEE lies in straightforwardly incinerating the waste material in furnaces. It is capital intensive, restricted to only high grade WEEE, emits hazardous gases, and is non-selective toward individual metals (Mäkinen et al., 2015). Moreover, smelting has caused many environmental problems associated to air pollution and heavy metal exposure in the past (Bakırdere et al., 2016). Biohydrometallurgy is an established technology for the extraction of metals from their primary sources, and currently more than 15% Cu and 5% of Au is produced by microorganisms (Johnson, 2014). Its application and mechanisms are well understood with the primary ores; however, its application to secondary ores is still at its infancy. WEEE, unlike primary metal sulfide ores, is a non-sulfide material, meaning that the discarded material does not include the minerals that provide an energy source for the bacteria. Thus, bioprocessing of WEEE requires externally added energy source to the bacteria. This applies both for the iron- and sulfur-oxidizer *Acidithiobacillus ferrivorans* and *Acidithiobacillus thiooxidans*, or cyanide-generating *Pseudomonas putida* (Chapter 5).

Selective recovery of metals from complex leachate solutions is an essential step. Typically, in a bioleaching leachate several metals are found in various concentrations in acidic sulfate medium along with metabolites resulting from the bioleaching reaction. The bioleaching leachate solution typically includes high concentrations of the bioleached metals from discarded PCB, such as Cu, Ni, Fe, Al, and Zn (Chapter 7). The main challenge is to achieve a selective recovery of metals with similar reactivity levels. Selective recovery of the metals from an aqueous leachate solution depends on several parameters such as the pH, concentration of the metals and their speciation.

9.2 Motivation for metal recovery from WEEE

Metal recovery from WEEE is primarily driven by the declining primary sources of these metals, coupled with the environmental burden of WEEE management. Modern devices encompass most metallic elements with an exponential increase of complexity of various mixtures of compounds (Chapter 4, Ongondo et al., 2015). The primary deposits of these metals are decreasing, many are projected to be insufficient to meet the current demand in a business as usual scenario (Graedel, 2011). In addition, their supply is at risk, due to uneven geographical distribution of their primary ores (Simoni et al., 2015). Their stable supply is essential for the well-being of the society and the transition to a sustainable, circular economy. In the tide of decoupling economic growth from hydrocarbon dependency, the circular economy is under risk of technology metals shortage.

In a circular economy, inclusion of secondary sources back into the economy through recycling and metal recovery is inevitable. In the emerging urban mining of end-of-life (EoL) devices the usage of waste as secondary raw materials plays a pivotal role. Inclusion of waste material back to the economy has several bottlenecks, including technological limitations, and low collection rates of the devices and poor enforcement of the laws concerning their management. These obstacles are interconnected, and amendment of one positively affects the other. In this dissertation, technical issues related to WEEE management are addressed and a product- and metal-specific recovery process is described. Specifically, a two-step procedure for the extraction and the recovery of Cu and Au from discarded PCB is designed, two potential metal extraction routes are investigated and their techno-economic and environmental sustainability assessment were assessed.

Hydrometallurgical processing is the norm to extract metals from their primary ores (Habashi, 2005). Also, many processes have been experimented to extract and recover metals from secondary sources, including WEEE (Tuncuk et al., 2012). Application of biological methods (biohydrometallurgy) is increasingly applied in the extraction of metals from primary sources since their introduction some 40 years ago. These biological processes are typically environmentally friendly, cost effective processes, where the pollutant production is minimal and process input is simply the nutrient requirement of the microorganisms. In addition, the waste source serves as a nutrient for microorganisms, in such waste-to-source processes. The mechanisms of bioprocessing of primary ores are relatively well-understood, however, there is a knowledge gap in bioprocessing of e-waste for metal recovery given the complexity of these materials and several metals, e.g. REE, are not present in the base metal ores.

The development of a novel metal recovery approach from waste materials is highly intricate, with a great number of factors involved. The performance of a metal recovery process mainly depends on the setting of the technical sub-system, the parameters, its efficiency to selectively recover metals from the waste material. Three major criteria determine the performance of the effectiveness of metals recovery from WEEE: **(1)** recovery efficiency of the selected technique, **(2)** selectivity toward individual metals, and **(3)** techno-economic and environmental profile of the technology. These criteria reflect the functional performance of the technology development. Scenarios with good performance in all these three criteria can establish the ultimate goal of the newly developed technology.

Research plays a major role in the systematic testing of techniques, understanding of the mechanisms, and evaluation of the performance of the technology in terms of effectiveness, and environmental profile. This research mainly focused the recovery efficiency, process parameters, selectivity towards individual metals, environmental profile, and techno-economic assessment of the technology. The performance was evaluated on the basis of the technical system. Evaluation of the base system can assist to identify the bottlenecks, flows and other gaps, and future research needs. This allows for further improvement of the technology in terms of effectiveness and environmental profile and brings it closer to the market. A strong scientific basis will lead to effective and sustainable resource recovery from electronic waste based on data gathering, analysis to determination of the technology.

9.3 Electronic waste as a secondary source of metals

9.3.1 Characterization of the discarded PCB from various sources

There is not yet a standardized method for the characterization of electronic waste material for total metal content. Many researchers reported different approaches, typically involving the digestion of the waste materials under extreme conditions (Park and Fray, 2009; Yang et al., 2011; Oguchi et al., 2012). A non-destructive total metal assay is typically carried out in a mixture of acids with an aims to extract the metals from the discarded material. A mixture of nitric and hydrochloric acid, a mixture known to dissolve Au and other valuable metals, is effective to achieve reliable and reproducible results from a polymetallic source (Melaku et al., 2005).

The modified characterization method given in Chapter 4 assayed total metals from various discarded PCB. The data obtained are in agreement with the reported values of other researchers who used similar approaches, in terms of dissolution method, and utilized analytical instrument. Many researchers used various digestive and non-digestive metal assay methods, typically involving nitro-hydrochloric acid in elevated temperatures. The microwave (MW) digestion method with nitro-hydrochloric acid showed practical significance as it is a rapid method, repeatable and is a standardizable procedure. Moreover, usage of corrosive hydrofluoric acid (HF) is avoided, in the total metal assay of a silicon-based material (PCB). However, it should be noted that nitro-hydrochloric digestion of crushed PCB might overlook the detection of some metals that react poorly with this mixture of acids. This might explain that some metals such as the lanthanides, silver and the other platinum group metals (PGM) were not detected (Chapter 4).

9.3.2 **Economic value and prioritization of metal recovery**

In urban mining of waste streams, WEEE, and specifically the PCB are an important and a strategic secondary source of valuable metals. The main economic motivation to recycle discarded PCB is the recovery of Cu and Au. These two metals make up to 92.3% to 97.8% of the total value (Chapter 4). PCB is a predominantly Cu-rich material (up to 38% by weight), which makes the recovery of this metal a priority in metal recovery operations. These findings are in agreement line with those of Wang and Gaustad (2012) who used statistical tools for the prioritization of metals. The selective recovery of Cu, the main material found in PCB, is a priority to serve a twofold purpose: to facilitate the recovery of other more valuable metals from PCB and to decrease a potential competition in the subsequent recovery step. Other metals are found in low quantities relative to Cu, in case of Au, a 8500-factor between Cu and Au was prominent in PCB from desktop computers (Chapter 4). Among other units of WEEE, PCB is the most valuable part owing to its high concentration of valuable metals (Chapter 2). An overview of major WEEE units, their metal concentration and the criticality of these metals are given in Table 9-1. The criticality information is taken from the Ad Hoc report of the European Commission on the critical metals (European Commission, 2014b) and available information from the literature (Jones et al., 2013; Rotter et al., 2013; Hennebel et al., 2015).

Table 9-1: Overview of secondary resources, their criticality and abundance.

WEEE unit	Critical metals	Criticality	Concentration compared to ores	Economic potential
Printed circuit boards (PCB)	Cu	Medium	High	Medium
	Au	Medium	Very high	Very high
Hard disc drives (HDD)	Nd	High	High	High
	Pr	High	High	High
Displays	In	High	High	Medium

9.3.3 **Factors influencing metal recovery technology development and application of multi-criteria assessment prior to research activities**

Technology selection is a cumbersome task given the variety of conventional and emerging technologies and the complexity of the polymetallic anthropogenic waste material. A novel approach is required to prioritize the metals of interest, and to develop an efficient metal recovery process. Several approaches to extract and subsequently recover metals are available, including conventional chemical and emerging biological technologies. The decision for technology selection should be based on an analytical ground, particularly when a number of multiple multidimensional criteria are involved. Multi-criteria analysis (MCA) using analytical hierarchical process (AHP) evaluated the available options for metal recovery technology, incorporating a number of relevant selection criteria at a hierarchical level of importance (Chapter 4).

9.4 Metal extraction from the discarded PCB

9.4.1 Biohydrometallurgy versus hydrometallurgy

Both chemical (hydrometallurgical) and biological (biohydrometallurgical) routes proved to be efficient for the extraction of Cu from discarded PCB. Bioleaching with a co-culture of *Acidithiobacillus ferrivorans* and *Acidithiobacillus thiooxidans* and chemical leaching with sulfuric acid (H_2SO_4) and hydrogen peroxide (H_2O_2) under optimized conditions resulted in a 98.4% and 99.2% Cu extraction efficiency, respectively (Chapters 5 and 6). However, in case of Au, chemical leaching (99.2%) is a lot more efficient than biological leaching to (44.0%), respectively. The biogenic cyanide production by the *Pseudomonas putida* was not high enough (21.6 mg/L), to leach the Au from the PCB, as was confirmed with chemical cyanide leaching tests (Chapter 5). In addition, rate there was a great difference in terms of dissolution, 21.3 and 205.7 factor multitude in favor of the chemical leaching route, respectively, for Cu and Au. An overview of bioleaching versus chemical leaching in terms of metals extraction efficiency and rate is given in Table 9-2.

Table 9-2: Comparison of leaching rate of biological and chemical routes.

Metals	Extraction efficiency (%) (Biological / Chemical)	Multitude	Leaching rate (µg /h) (Biological / Chemical)	Multitude
Cu	98.4 / 99.2	1.008	172 / 367	21.3
Au	44.0 / 96.2	2.18	0.192 / 39.5	205.7

A two-step leaching procedure was developed to separate the extraction process for the base and precious metals, i.e. Cu and Au found in the discarded PCB. Cu competes with Au in metal extraction process and increases the leachant consumption, particularly in a multi-metal source such as PCB, where both metals are found in very high concentrations. Several researchers developed strategies to enhance the biogenic cyanide production (Natarajan et al., 2015b) and/or chemical stability of cyanide in solution (Natarajan and Ting, 2015a). In its current status, bioleaching of Au has a lot to improve in order to compete with chemical technologies.

It is worth to mention that all bioleaching experiments were carried out at a low technology readiness level (TRL) in batch processes in agitated flasks. The reaction rates could differ greatly at full scale, when many other factors are included. At an industrial scale, bioleaching of primary ores is carried out in large heaps or in controlled stirred vats. From a process engineering point view, a continuous stirring tank reactor (CSTR) would be the most appropriate selection to control the complex bioleaching process (Acevedo, 2000). A scale-up procedure and sensitivity analysis is essential, particularly for the biomass-based technologies. In a scale-up bioleaching testing system, i.e. a semi-pilot scale continuous bioleaching reactor, many contribute to the enhancement of bioleaching rate and efficiency. For instance, in a bioleaching reactor, the bacteria are enhanced with sparging O_2 and CO_2, which have an immense effect on the bioleaching activity of acidophiles (Brierley and Brierley, 2013). Recently, many researchers proved a substantially increased bioleaching rate (Ilyas and Lee, 2014b; Mäkinen et al., 2015) under enhanced optimized conditions (Vera et al., 2013) or combined with chemical giving rise to hybrid technologies (Ilyas et al., 2015). The process is not necessarily restricted to thermophilic leaching and can be carried out under mesophilic conditions (Chen et al., 2015; Chapter 5) and the bacteria can be loaded with a high pulp density up to 10% (Ilyas and Lee, 2014b).

Following the developments in the early 2000s in the fundamentals of biomining of metals from minerals (Rohwerder et al., 2003), biohydrometallurgical routes succeeded from uncontrolled waste heaps to engineered tank leaching (Watling, 2015). Bioleaching of non-sulfide anthropogenic materials has an unexplored potential in terms of leaching rate, higher waste load, and unexplored microbial species. Further improvement of these parameters can make bioleaching of base metals to be competitive as chemical technologies. In this context, bioprocessing of WEEE for metal recovery is a field open to many new developments. On the other hand, hydrometallurgical processes, the limits of the processes are already mostly outlined, and the stoichiometry of the leaching reactions is rigid, and the limits of hydrometallurgical processes are well studied.

9.4.2 Cyclic bioleaching reaction and iron speciation

A cyclic Fe^{2+} - Fe^{3+} reaction was observed in Cu bioleaching from PCB (Chapter 5). This cyclic reaction of Fe gives also an opportunity to recirculate the solution back to the bioleaching step, where Fe^{2+} can serve as an electron donor for the bioleaching bacteria. Bioleaching of base metals from WEEE involved the iron-oxidizer *Acidithiobacillus ferrivorans*, which required the external addition of ferrous iron (Fe^{2+}) to the bioleaching medium. The bacterial Fe^{2+} oxidation is given below in Equation (9-1:

$$4Fe^{2+} + O_2 + 2H^+ \rightarrow 4Fe^{3+} + 2OH^- \qquad (9\text{-}1)$$

Biogenic ferric (Fe^{3+}) acts as a leachant for the bioleaching of Cu as shown below in Equation (9-2. In turn, Fe^{3+} is reduced as a result of the redox reaction with the elemental Cu found in the waste material (Chapter 5).

$$Cu^0 + 2Fe^{3+} \rightarrow Cu^{2+} + 2Fe^{2+} \qquad (9\text{-}2)$$

The bioleaching leachate solution was found to contain predominantly (96%) Fe^{2+}, when all the Cu was leached from the waste material, at the point where the bioleaching reaction was ceased and all the bacterial cells were removed from the solution (Chapter 6). This gives an opportunity to use the bioleaching solution as a feed for the first bioleaching reaction. Moreover, the selective recovery of Cu from the bioleaching leachate solution was successful owing to the predominantly ferrous dominated leachate solution. An illustrative schematic is given in Figure 9-1.

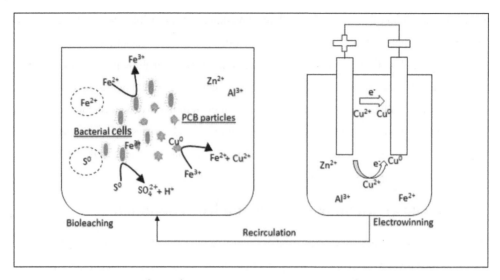

Figure 9-1: Cyclic iron (Fe^{2+} - Fe^{3+}) reaction and recirculation of Fe^{2+}-containing solution into the bioleaching solution.

9.5 Recovery: electrowinning versus sulfidic precipitation

Both electrowinning (EW) and sulfidic precipitation are efficient to selectively recovering Cu (98.4% / 91.2%) from a complex bioleaching solution (Chapter 7). The bioleaching solution contained a relative high organic content (38.2 mg/L total organic carbon) and a complex mixture of metals, dominated primarily by iron (Fe) and Cu (Chapter 7). Both recovery techniques were selective towards Cu, producing a final product predominantly containing Cu, with a purity of 65% and 35%, respectively, with EW and precipitation. Further studies on the process parameters, that complete selective recovery of Cu from the bioleaching solution is possible, when the process parameters are optimized, for these techniques.

Sulfidic precipitation of Cu was stoichiometric at the pH value (pH = 1.05) of the bioleaching solution. At various sulfide concentrations (1 – 100 mM). Cu was selectively precipitated over other cations (Fe, Zn, Ni, Al), due to the difference in solubility constant between the metal-sulfide products. At stoichiometric concentrations of S^{2-}, 86% of the Cu was precipitated from the solution. Further investigations showed that the precipitate was a covellite mineral (CuS) including 35% Cu and 35% S (by weight), along with minor fractions of Fe as well (Chapter 5). The particle size of the covellite precipitates were 116 nm in average with good settleability properties.

EW was applied without solvent extraction in our tests (Chapter), and Fe was recovered by the application of low current (50 mA). This created an opportunity to recover Cu by applying a low current, without the requirement of a complexing agent. Most commonly in the leachate solutions, Fe was found in its oxidized ferric (Fe^{3+}) form and it competes with Cu in the electrodeposition system, thus requiring the addition of a complexing agent. These findings were confirmed with the experiments carried out with a synthetic leachate solution, in which ferrous iron (Fe^{2+}) was used as the sole Fe source. The speciation of Fe is a key element in this system. This element is typically found predominantly in its Fe^{3+} in the leachates from primary ores. In bioleaching tests, however, Fe was predominantly found in ferrous form Fe^{2+}. This finding showed that the cyclic reaction of Fe played an important role in bioprocessing of waste material using iron-oxidizers in an integrated (bioleaching + electrowinning recovery) system. Moreover, Cu was selectively recovered over Fe, because it was found in its Fe^{2+} form, a species much less reactive than the oxidized Fe^{3+} form. However, when the solution was predominantly Fe^{2+}, a competition was not prevalent between Cu and Fe and a low current was sufficient to selectively deposit Cu on the electrode and leave ferrous-dominated Fe fraction in the solution.

9.6 Techno-economic assessment and environmental sustainability analysis of an emerging technology at an early stage of development

Economic and environmental factors are decisive in the development of full-scale industrial processes. The operational costs are the decisive factor in selection between alternative routes. Moreover, they play an integral role in the design of the process route, the tank and the throughput rate, thus indirectly affecting the process economics.

It should be noted that primary laboratory data was used in the scale-up calculation and to the greatest extent possible (Chapter 8). Besides the sensitivity analysis of the primary data, many uncertainties exist which are not incorporated in the scale-up, such as macro-economic factors, metal price volatility, and costs that associate with the operational costs of thermal recovery process. The calculations were carried out in very optimistic scenarios, in which high grade PCB (high Cu concertation, high Au concertation) was modelled. The Au content is the main factor, contributing largely to the total revenues. The potential values of the other metals were not modelled, as they were not detected, or their economic value was insignificant (Chapter 4).

A comparative environmental sustainability analysis was carried out with the goal of comparing the alternative processes, and defining the environmental hotspots in each step, rather than obtaining absolute results that can be benchmarked to existing full scale applications. Energy consumption and its related hazards was the main source in every process alternative, namely biological processing, chemical processing and hybrid processing.

Biological processing of primary ores has great advantages over chemical processing, owing to less impactful effects on the climate, elimination of hazardous chemicals (Panda et al., 2015), However in bioprocessing of non-sulfidic, secondary raw materials, such WEEE, the bacteria need to be externally supplemented (Chapter 5), which brings along additional environmental impacts (Chapter 8). Lastly, it is worth to mention that the life cycle assessment (LCA) assessment was not compared to the production of the metals from their primary ores (Chapter 8), on which many researchers showed that great natural resource preservation and energy savings, and climate benefits can be gained (Liu et al., 2009).

9.7 Strategies for the development of a sustainable technology to recover metals from electronic waste

Current best available technologies (BAT) stimulates the recovery of metals through pyrometallurgical routes on the basis of thermal treatment of the discarded material. Currently, a number of smelters operate to recover valuable metals from WEEE in high temperature smelters. It is, however, limited to only high grade WEEE, and typically solely PCB with a high precious metal content can be treated economically (Schluep et al., 2009; Wellmer and Hagelüken, 2015). The concentration of the precious metals, and mostly notably Au, is expected to decrease owing to changing manufacturing technologies and emerging ultra-thin coating technology, down to a few atom thick contact layer in the modern devices (Luda, 2011; Birloaga et al., 2013; Akcil et al., 2015).

On the other hand, as the complexity of the devices increases, particularly in the newest devices, a higher number of elements are found in EoL EEE in smaller amounts and higher complexity levels. Therefore, selectivity towards individual metals is the key priority in metal recovery from WEEE. Critical metal recovery from this urban waste stream might prove inefficient through pyrometallurgical routes, given the chemical properties of these metals. Therefore, technologies addressing selectivity toward individual metals need to be developed for the future of metal recovery from WEEE. In this regard, hydrometallurgical, and biohydrometallurgical processing might have advantages over pyrometallurgical processing. On the other hand, pyrometallurgical processing might be upgraded to more environmentally friendly processes, by limiting its hazardous gaseous emissions and energy consumption. Most likely, a combination of these techniques in highly complex processes will be required to selectively and sustainably recover metals from various streams of polymetallic WEEE. In synthesis, sustainable recovery of metals from secondary sources requires a multidisciplinary approach, requiring inputs from environmental engineering, hydrometallurgy, solid waste management, and industrial ecology.

9.8 Overall conclusions

- WEEE and particularly discarded PCB are an important secondary source of metals. Cu and Au are very valuable and their efficient recovery from discarded PCB should be prioritized. It is expected that the discarded PCB will be a major source of these metals in the coming decade.

- Recovery technologies are at their infancy, and most industrial scale applications are limited to pyrometallurgical routes applied to a small fraction of metal-rich WEEE. In its current status, this technology is limited to high grade PCB with a high valuable metal content. Thus, development of metal-and product-specific technologies are required.

- Discarded PCB is a very complex material in terms of metal abundance, concentration, and chemical structure. Unlike primary ores, the metals are found in their metallic forms, often in combination with alloys. Cu and Au are the most valuable metals and the uttermost priority to be recovered.

- Hydrometallurgy and biohydrometallurgy are two viable alternatives. Hydrometallurgy has relatively more robust processes, faster kinetics, reliable application and higher yield. Biohydrometallurgy is at its infancy, despite the extensive research efforts and sound understanding of fundamental concepts on its application on primary ores. Following the success of biohydrometallurgy with the primary ores in the last two decades, and its improvement potential, new developments can be expected in the near-future. In metal extraction from PCB, or other polymetallic sources, a hybrid approach, using the advantages of biohydrometallurgical and hydrometallurgical processing might prove to be the best option.

References

1. Abbruzzese C, Fornari P, Massidda R et al. (1995) Thiosulphate leaching for gold hydrometallurgy. Hydrometallurgy 39:265–276. doi:10.1016/0304-386X(95)00035-F.

2. Acevedo F (2000) The use of reactors in biomining processes. Electron J Biotechnol 3:184–194. doi:10.2225/vol3-issue3-fulltext-4.

3. Ahluwalia PK, Nema AK (2007) A life cycle based multi-objective optimization model for the management of computer waste. Resour Conserv Recycl 51:792–826. doi:10.1016/j.resconrec.2007.01.001.

4. Ahmad FB, Zhang Z, Doherty WOS, O'Hara IM (2015) A multi-criteria analysis approach for ranking and selection of microorganisms for the production of oils for biodiesel production. Bioresour Technol 190:264–273. doi:10.1016/j.biortech.2015.04.083.

5. Akcil A (2010) A new global approach of cyanide management: International cyanide management code for the manufacture, transport, and use of cyanide in the production of gold. Miner Process Extr Metall Rev 31:135–149.

6. Akcil A, Erust C, Gahan CS et al. (2015) Precious metal recovery from waste printed circuit boards using cyanide and non-cyanide lixiviants - A review. Waste Manag 45:258–271. doi:10.1016/j.wasman.2015.01.017.

7. Andrès Y, Gérente C (2011) Removal of rare earth elements and precious metal species by biosorption. In: Kotrba P, Mackova M, Macek T (eds) Microb. biosorption Met. Springer Netherlands, Dordrecht, pp 179–196.

8. Antonopoulos IS, Perkoulidis G, Logothetis D, Karkanias C (2014) Ranking municipal solid waste treatment alternatives considering sustainability criteria using the analytical hierarchical process tool. Resour Conserv Recycl 86:149–159. doi:10.1016/j.resconrec.2014.03.002.

9. Arundel A, Sawaya D (2009) The bioeconomy to 2030: Desining a policy agenda. OECD Publishing.

10. Auernik KS, Maezato Y, Blum PH, Kelly RM (2008) The genome sequence of the metal-mobilizing, extremely thermoacidophilic archaeon *Metallosphaera sedula* provides insights into bioleaching-associated metabolism. Appl Environ Microbiol 74:682–692. doi:10.1128/AEM.02019-07.

11. Aylmore MG, Muir DM (2001) Thiosulfate leaching of gold—A review. Miner Eng 14:135–174. doi:10.1016/S0892-6875(00)00172-2.

12. Azizi D, Shafaei SZ, Noaparast M, Abdollahi H (2012) Modeling and optimization of low-grade Mn bearing ore leaching using response surface methodology and central composite rotatable design. Trans Nonferrous Met Soc China (English Ed 22:2295–2305. doi:10.1016/S1003-6326(11)61463-5.

13. Baird J, Curry R, Cruz P (2014) An overview of waste crime, its characteristics, and the vulnerability of the EU waste sector. Waste Manag Res 32:97–105. doi:10.1177/0734242X13517161.

14. Bakas I, Fischer C, Harding A (2014) Present and potential future recycling of critical metals in WEEE. Copenhagen Resour. Inst.

15. Bakırdere S, Bölücek C, Yaman M (2016) Determination of contamination levels of Pb, Cd, Cu, Ni, and Mn caused by former lead mining gallery. Environ Monit Assess 188:1–7. doi:10.1007/s10661-016-5134-5.

16. Bakker PAHM, Pieterse CMJ, Loon LC Van (2007) Induced systemic resistance by fluorescent *Pseudomonas* spp . 97:239–243.

17. Baldé CP, Wang F, Kuehr R, Huisman J (2015) The global e-waste monitor 2014, United Nationas University, IAS - SCYCLE, Bonn.

18. Bas AD, Deveci H, Yazici EY (2013) Bioleaching of copper from low grade scrap TV. circuit boards using mesophilic bacteria. Hydrometallurgy 138:65–70. doi:10.1016/j.hydromet.2013.06.015.

19. Batstone DJ, Hülsen T, Mehta CM, Keller J (2015) Platforms for energy and nutrient recovery from domestic wastewater: A review. Chemosphere 140:2–11. doi:10.1016/j.chemosphere.2014.10.021.

20. Bebelis S, Bouzek K, Cornell A et al. (2013) Highlights during the development of electrochemical engineering. Chem Eng Res Des 91:1998–2020. doi:10.1016/j.cherd.2013.08.029.

21. Behnamfard A, Salarirad MM, Veglio F (2013) Process development for recovery of copper and precious metals from waste printed circuit boards with emphasize on palladium and gold leaching and precipitation. Waste Manag 33:2354–2363. doi:10.1016/j.wasman.2013.07.017.

22. Bharadwaj A, Ting Y (2011) From biomining of mineral ores to bio urban mining of industrial waste. Environ. Technol. Manag. Conf. 4th ETMC.

23. Biddy MJ, Davis R, Humbird D et al. (2016) The techno-economic basis for coproduct manufacturing to enable hydrocarbon fuel production from lignocellulosic biomass. ACS Sustain Chem Eng 4:3196–3211. doi:10.1021/acssuschemeng.6b00243.

24. Bigum M, Brogaard L, Christensen TH (2012) Metal recovery from high-grade WEEE: A life cycle assessment. J Hazard Mater 207–208:8–14. doi:10.1016/j.jhazmat.2011.10.001.

25. Birloaga I, Coman V, Kopacek B, Vegliò F (2014) An advanced study on the hydrometallurgical processing of waste computer printed circuit boards to extract their valuable content of metals. Waste Manag 34:2581–2586. doi:10.1016/j.wasman.2014.08.028.

26. Birloaga I, De Michelis I, Ferella F et al. (2013) Study on the influence of various factors in the hydrometallurgical processing of waste printed circuit boards for copper and gold recovery. Waste Manag 33:935–941. doi:10.1016/j.wasman.2013.01.003.

27. Biswas S, Chakraborty S, Chaudhuri MG et al. (2014) Optimization of process parameters and dissolution kinetics of nickel and cobalt from lateritic chromite overburden using organic acids. J Chem Techology Biotechnol 89:1491–1500. doi:10.1002/jctb.4288.

28. Bloodworth A (2014) Track flows to manage technology-metal supply. Nature 505:9–10. doi:10.1038/505019a.

29. Blumer C, Haas D (2000) Mechanism, regulation, and ecological role of bacterial cyanide biosynthesis. Arch Microbiol 173:170–7.

30. Bolzán AE (2013) Electrodeposition of copper on glassy carbon electrodes in the presence of picolinic acid. Electrochim Acta 113:706–718. doi:10.1016/j.electacta.2013.09.132.

31. Bonnefoy V, Holmes DS (2012) Genomic insights into microbial iron oxidation and iron uptake strategies in extremely acidic environments. Environ Microbiol 14:1597–1611. doi:10.1111/j.1462-2920.2011.02626.x.

32. Bosecker K (1997) Bioleaching: Metal solubilization by microorganisms. FEMS Microbiol Rev 20:591–604. doi:10.1016/S0168-6445(97)00036-3.

33. Brandl H (2008) Microbial leaching of metals. Biotechnol. Set, Second Ed. pp 191–206.

34. Brandl H, Bosshard R, Wegmann M (2001) Computer-munching microbes: metal leaching from electronic scrap by bacteria and fungi. Hydrometallurgy 59:319–326. doi:10.1016/S0304-386X(00)00188-2.

35. Brandl H, Faramarzi M (2006) Microbe-metal-interactions for the biotechnological treatment of metal-containing solid waste. China Particuology 4:93–97. doi:10.1016/S1672-2515(07)60244-9.

36. Brandl H, Lehmann S, Faramarzi MA, Martinelli D (2008) Biomobilization of silver, gold, and platinum from solid waste materials by HCN-forming microorganisms. Hydrometallurgy 94:14–17. doi:10.1016/j.hydromet.2008.05.016.

37. Brasseur G, Levican G, Bonnefoy V et al. (2004) Apparent redundancy of electron transfer pathways via bc1 complexes and terminal oxidases in the extremophilic chemolithoautotrophic *Acidithiobacillus ferrooxidans*. Biochim Biophys Acta - Bioenerg 1656:114–126. doi:10.1016/j.bbabio.2004.02.008.

38. Breivik K, Armitage JM, Wania F, Jones KC (2014) Tracking the global generation and exports of e-waste. Do existing estimates add up? Environ Sci Technol 48:8735–8743. doi:10.1021/es5021313.

39. Breuer PL, Jeffrey MI (2000) Thiosulfate leaching kinetics of gold in the presence of copper and ammonia. Miner Eng 13:1071–1081. doi:10.1016/S0892-6875(00)00091-1.

40. Brierley CL, Brierley J a. (2013) Progress in bioleaching: Part B: Applications of microbial processes by the minerals industries. Appl Microbiol Biotechnol 97:7543–7552. doi:10.1007/s00253-013-5095-3.

41. Brierley J., Brierley C. (2001) Present and future commercial applications of biohydrometallurgy. Hydrometallurgy 59:233–239. doi:10.1016/S0304-386X(00)00162-6.

42. Brierley JA, Brierley CL (1986) Microbial mining using thermophilic microorganisms. In: Brock TD (ed) Thermophiles Gen. Mol. Appl. Microbiol. John Wiley & Sons, New York, pp 279–305.

43. Bryan CG, Watkin EL, McCredden TJ et al. (2015) The use of pyrite as a source of lixiviant in the bioleaching of electronic waste. Hydrometallurgy 152:33–43. doi:10.1016/j.hydromet.2014.12.004.

44. Buchert M, Manhart A, Bleher D, Pingel D (2012) Recycling kritischer Rohstoffe aus Elektronik-Altgeräten. LANUV-Fachbericht 38. Freiburg, Germany.

45. Campbell SC, Olson GJ, Clark TR, McFeters G (2001) Biogenic production of cyanide and its application to gold recovery. J Ind Microbiol Biotechnol 26:134–139. doi:10.1038/sj.jim.7000104.

46. Canovas D, Cases I, de Lorenzo V (2003) Heavy metal tolerance and metal homeostasis in *Pseudomonas putida* as revealed by complete genome analysis. Environ Microbiol 5:1242–1256. doi:10.1046/j.1462-2920.2003.00463.x.

47. Cao J, Zhang G, Mao Z et al. (2009) Precipitation of valuable metals from bioleaching solution by biogenic sulfides. Miner Eng 22:289–295. doi:10.1016/j.mineng.2008.08.006.

48. Cappuyns V, Swennen R, Vandamme A, Niclaes M (2006) Environmental impact of the former Pb-Zn mining and smelting in East Belgium. J Geochemical Explor 88:6–9. doi:10.1016/j.gexplo.2005.08.005.

49. Carn SA, Krueger AJ, Krotkov NA et al. (2007) Sulfur dioxide emissions from Peruvian copper smelters detected by the ozone monitoring instrument. Geophys. Res. Lett. 34: L09801. doi:10.1029/2006GL029020.

50. Castric PA (1977) Glycine metabolism by *Pseudomonas aeruginosa*: hydrogen cyanide biosynthesis. J Bacteriol 130:826–831.

51. Chancerel P, Bolland T, Rotter VS (2011) Status of pre-processing of waste electrical and electronic equipment in Germany and its influence on the recovery of gold. Waste Manag Res 29:309–17. doi:10.1177/0734242X10368303.

52. Chancerel P, Meskers CEM, Hagelüken C, Rotter VS (2009) Assessment of precious metal flows during preprocessing of waste electrical and electronic equipment. J Ind Ecol 13:791–810. doi:10.1111/j.1530-9290.2009.00171.x.

53. Chancerel P, Rotter S (2009) Recycling-oriented characterization of small waste electrical and electronic equipment. Waste Manag 29:2336–2352. doi:10.1016/j.wasman.2009.04.003.

54. Chang D, Lee CKM, Chen C-H (2014) Review of Life Cycle Assessment towards Sustainable Product Development. J Clean Prod 83:48–60. doi:10.1016/j.jclepro.2014.07.050.

55. Chen S, Yang Y, Liu C et al. (2015) Column bioleaching copper and its kinetics of waste printed circuit boards (WPCBs) by *Acidithiobacillus ferrooxidans*. Chemosphere 141:162–168. doi:10.1016/j.chemosphere.2015.06.082.

56. Chen SY, Huang QY (2014) Heavy metals recovery from printed circuit board industry wastewater sludge by thermophilic bioleaching process. J Chem Technol Biotechnol 89:158–164. doi:10.1002/jctb.4129.

57. Chen T, Lei C, Yan B, Xiao X (2014) Metal recovery from the copper sulfide tailing with leaching and fractional precipitation technology. Hydrometallurgy 147–148:178–182. doi:10.1016/j.hydromet.2014.05.018.

58. Chi T, Lee J, Pandey BD et al. (2011) Bioleaching of gold and copper from waste mobile phone PCBs by using a cyanogenic bacterium. Miner Eng 24:1219–1222. doi:10.1016/j.mineng.2011.05.009.

59. Chung J, Jeong E, Choi JW et al. (2015) Factors affecting crystallization of copper sulfide in fed-batch fluidized bed reactor. Hydrometallurgy 152:107–112. doi:10.1016/j.hydromet.2014.12.01.4.

60. Cimpan C, Maul A, Jansen M et al. (2015) Central sorting and recovery of MSW recyclable materials: A review of technological state-of-the-art, cases, practice and implications for materials recycling. J Environ Manage 156:181–199. doi:10.1016/j.jenvman.2015.03.025.

61. Clark DA, Norris PR (1996) *Acidimicrobium ferrooxidans* gen nov, sp nov: Mixed-culture ferrous iron oxidation with Sulfobacillus species. Microbiology 142:785–790. doi:10.1099/00221287-142-4-785.

62. Colmer AR, Hinkle ME (1947) The role of microorganisms in acid mine drainage: A preliminary report. Science 106:253–6. doi:10.1126/science.106.2751.253.

63. Corstjens PLAM, De Vrind JPM, Westbroek P, De Vrind-De Jong EW (1992) Enzymatic iron oxidation by Leptothrix discophora: Identification of an iron-oxidizing protein. Appl Environ Microbiol 58:450–454.

64. Creamer NJ, Baxter-Plant VS, Henderson J et al. (2006) Palladium and gold removal and recovery from precious metal solutions and electronic scrap leachates by *Desulfovibrio desulfuricans*. Biotechnol Lett 28:1475–1484. doi:10.1007/s10529-006-9120-9.

65. Cucchiella F, D'Adamo I, Gastaldi M (2014) Sustainable management of waste-to-energy facilities. Renew Sustain Energy Rev 33:719–728. doi:10.1016/j.rser.2014.02.015.

66. Cucchiella F, D'Adamo I, Lenny Koh SC, Rosa P (2015) Recycling of WEEEs: An economic assessment of present and future e-waste streams. Renew Sustain Energy Rev 51:263–272. doi:10.1016/j.rser.2015.06.010.

67. Cui J, Forssberg E (2003) Mechanical recycling of waste electric and electronic equipment: A review. J Hazard Mater 99:243–263. doi:10.1016/S0304-3894(03)00061-X.

68. Cui J, Forssberg E (2007) Characterization of shredded television scrap and implications for materials recovery. Waste Manag 27:415–424. doi:10.1016/j.wasman.2006.02.003.

69. Cui J, Zhang L (2008) Metallurgical recovery of metals from electronic waste: A review. J Hazard Mater 158:228–256. doi:10.1016/j.jhazmat.2008.02.001.

70. Cundy AB, Bardos RP, Church A et al. (2013) Developing principles of sustainability and stakeholder engagement for "gentle" remediation approaches: The European context. J Environ Manage 129:283–291. doi:10.1016/j.jenvman.2013.07.032.

71. Dai X, Breuer PL (2013) Leaching and electrochemistry of gold, silver and gold–silver alloys in cyanide solutions: Effect of oxidant and lead(II) ions. Hydrometallurgy 133:139–148. doi:10.1016/j.hydromet.2013.01.002.

72. Dang MT, Brunner P-LM, Wuest JD (2014) A green approach to organic thin-film electronic devices: Recycling electrodes composed of Indium Tin Oxide (ITO). Sustain Chem Eng 2:2715–2721. doi:10.1021/sc500456p.

73. Darland G, Brock TD, Samsonoff W, Conti SF (1970) A thermophilic, acidophilic mycoplasma isolated from a coal refuse pile. Science 80, 170:1416–1418. doi:10.1126/science.170.3965.1416.

74. Darnall DW, Greene B, Henzl MT et al. (1986) Selective recovery of gold and other metal ions from an algal biomass. Environ Sci Technol 20:206–208.

75. Das N (2010) Recovery of precious metals through biosorption - A review. Hydrometallurgy 103:180–189. doi:10.1016/j.hydromet.2010.03.016.

76. Das SC, Gopala Krishna P (1996) Effect of Fe(III) during copper electrowinning at higher current density. Int J Miner Process 46:91–105. doi:10.1016/0301-7516(95)00056-9.

77. Deng X, Chai L, Yang Z et al. (2013) Bioleaching mechanism of heavy metals in the mixture of contaminated soil and slag by using indigenous *Penicillium chrysogenum* strain F1. J Hazard Mater 248–249:107–114. doi:10.1016/j.jhazmat.2012.12.051.

78. Deplanche K, Macaskie LE (2008) Biorecovery of gold by *Escherichia coli* and *Desulfovibrio desulfuricans*. Biotechnol Bioeng 99:1055–1064. doi:10.1002/bit.21688.

79. Deplanche K, Murray A, Mennan C et al. (2005) Biorecycling of precious metals and rare earth elements.

80. Dewulf J, Mancini L, Blengini GA et al. (2015) Toward an overall analytical framework for the integrated sustainability assessment of the production and supply of raw materials and primary energy carriers. J Ind Ecol 19:963–977. doi:10.1111/jiec.12289.

81. Dimitrov N (2016) Recent advances in the growth of metals, alloys, and multilayers by surface limited redox replacement (SLRR) based approaches. Electrochim Acta 209:599–622. doi:10.1016/j.electacta.2016.05.115.

82. Donati ER, Sand W (2007) Microbial processing of metal sulfides. Springer.

83. Dreisinger D (2006) Copper leaching from primary sulfides: Options for biological and chemical extraction of copper. Hydrometallurgy 83:10–20. doi:10.1016/j.hydromet.2006.03.032.

84. Duan H, Hou K, Li J, Zhu X (2011) Examining the technology acceptance for dismantling of waste printed circuit boards in light of recycling and environmental concerns. J Environ Manage 92:392–399. doi:10.1016/j.jenvman.2010.10.057.

85. Edwards KJ, Bond PL, Gihring TM, Banfield JF (2000) An archaeal iron-oxidizing extreme acidophile important in acid mine drainage. Science 287:1796–1799. doi:10.1126/science.287.5459.1796.

86. Erüst C, Akcil A, Gahan CS et al. (2013) Biohydrometallurgy of secondary metal resources: A potential alternative approach for metal recovery. J Chem Technol Biotechnol 88:2115–2132. doi:10.1002/jctb.4164.

87. Escobar B, Jedlicki E, Wiertz J, Vargas T (1996) A method for evaluating the proportion of free and attached bacteria in the bioleaching of chalcopyrite with *Thiobacillus ferrooxidans*. Hydrometallurgy 40:1–10. doi:10.1016/0304-386X(95)00005-2.

88. European Commission (2014) Report on critical raw materials for the EU, Report of the Ad hoc Working Group on defining critical raw materials.

89. Eurostat (2016) Eurostat. In: Netherlands energy prices. http://ec.europa.eu/eurostat/statistics-explained/index.php/Energy_price_statistics. Accessed 1 Sep 2016.

90. Van Eygen E, De Meester S, Tran HP, Dewulf J (2016) Resource savings by urban mining: The case of desktop and laptop computers in Belgium. Resour Conserv Recycl 107:53–64. doi:10.1016/j.resconrec.2015.10.032.

91. Fabisch M, Beulig F, Akob DM, Küsel K (2013) Surprising abundance of *Gallionella*-related iron oxidizers in creek sediments at pH 4.4 or at high heavy metal concentrations. Front Microbiol 4:1–12. doi:10.3389/fmicb.2013.00390.

92. Fairbrother L, Shapter J, Brugger J et al. (2009) Effect of the cyanide-producing bacterium Chromobacterium violaceum on ultraflat Au surfaces. Chem Geol 265:313–320. doi:10.1016/j.chemgeo.2009.04.010.

93. Fan SKS, Fan C, Yang JH, Liu KFR (2013) Disassembly and recycling cost analysis of waste notebook and the efficiency improvement by re-design process. J Clean Prod 39:209–219. doi:10.1016/j.jclepro.2012.08.014.

94. Faramarzi M, Stagars M, Pensini E (2004) Metal solubilization from metal-containing solid materials by cyanogenic *Chromobacterium violaceum*. J Biotechnol 113:321–326. doi:10.1016/j.jbiotec.2004.03.031.

95. Ficeriová J, Baláž P, Gock E (2011) Leaching of gold, silver and accompanying metals from circuit boards (PCBs) waste. Acta Montan Slovaca 16:128–131.

96. Fogarasi S, Imre-Lucaci F, Imre-Lucaci Á, Ilea P (2014) Copper recovery and gold enrichment from waste printed circuit boards by mediated electrochemical oxidation. J Hazard Mater 273:215–221. doi:10.1016/j.jhazmat.2014.03.043.

97. Foulkes JM, Deplanche K, Sargent F et al. (2016) A Novel Aerobic Mechanism for Reductive Palladium Biomineralization and Recovery by *Escherichia coli*. Geomicrobiol J 33:230–236. doi:10.1080/01490451.2015.1069911.

98. Fowler PW, Orwick-Rydmark M, Radestock S et al. (2015) Gating Topology of the Proton-Coupled Oligopeptide Symporters. Structure 23:290–301. doi:10.1016/j.str.2014.12.012.

99. Friege H (2012) Review of material recovery from used electric and electronic equipment-alternative options for resource conservation. Waste Manag Res 30:3–16. doi:10.1177/0734242X12448521.

100. Frischknecht R, Büsser S, Krewitt W (2009) Environmental assessment of future technologies: How to trim LCA to fit this goal? Int J Life Cycle Assess 14:584–588. doi:10.1007/s11367-009-0120-6.

101. Fu F, Wang Q (2011) Removal of heavy metal ions from wastewaters: A review. J Environ Manage 92:407–418. doi:10.1016/j.jenvman.2010.11.011.

102. Gadd GM (2010) Metals, minerals and microbes: Geomicrobiology and bioremediation. Microbiology 156:609–643. doi:10.1099/mic.0.037143-0.

103. Gargalo CL, Carvalho A, Gernaey K V., Sin G (2016) A framework for techno-economic & environmental sustainability analysis by risk assessment for conceptual process evaluation. Biochem. Eng. J. http://dx.doi.org/10.1016/j.bej.2016.06.007.

104. Gavankar S, Suh S, Keller AA (2015) The role of scale and technology maturity in Life Cycle Assessment of emerging technologies: A Case Study on Carbon Nanotubes. J Ind Ecol 19:51–60. doi:10.1111/jiec.12175.

105. Georgiadis DR, Mazzuchi TA, Sarkani S (2013) Analysis of alternatives for selection of enabling technology. Syst Eng 16:287–303. doi:DOI 10.1002/sys.

106. Ghauri MA, Okibe N, Barrie Johnson D (2007) Attachment of acidophilic bacteria to solid surfaces: The significance of species and strain variations. Hydrometallurgy 85:72–80. doi:10.1016/j.hydromet.2006.03.016.

107. Ghodrat M, Rhamdhani MA, Brooks G et al. (2016) Techno economic analysis of electronic waste processing through black copper smelting route. J Clean Prod 126:178–190. doi:10.1016/j.jclepro.2016.03.033.

108. Ghosh B, Ghosh MK, Parhi P et al. (2015) Waste Printed Circuit Boards recycling: An extensive assessment of current status. J Clean Prod 94:5–19. doi:10.1016/j.jclepro.2015.02.024.

109. Golyshina O V., Pivovarova TA, Karavaiko GI et al. (2000) *Ferroplasma acidiphilum* gen. nov., sp. nov., an acidophilic, autotrophic, ferrous-iron-oxidizing, cell-wall-lacking, mesophilic member of the Ferroplasmaceae fam. nov., comprising a distinct lineage of the Archaea. Int J Syst Evol Microbiol 50:997–1006. doi:10.1099/00207713-50-3-997.

110.   González-Muñoz MJ, Rodríguez MA, Luque S, Álvarez JR (2006) Recovery of heavy metals from metal industry waste waters by chemical precipitation and nanofiltration. Desalination 200:742–744. doi:10.1016/j.desal.2006.03.498.

111.   Gorgievski M, Božić D, Stanković V, Bogdanović G (2009) Copper electrowinning from acid mine drainage: A case study from the closed mine "Cerovo." J Hazard Mater 170:716–721. doi:10.1016/j.jhazmat.2009.04.135.

112.   Graedel TE (2011) Metal Stocks & Recycling Rates. UNEP, International Resource Panel. ISBN: 978-92-807-3182-0.

113.   Graedel TE, Erdmann L (2012) Will metal scarcity impede routine industrial use? MRS Bull 37:325–331. doi:10.1557/mrs.2012.34.

114.   Grosse AC, Dicinoski GW, Shaw MJ, Haddad PR (2003) Leaching and recovery of gold using ammoniacal thiosulfate leach liquors (a review). Hydrometallurgy 69:1–21. doi:10.1016/S0304-386X(02)00169-X.

115.   Gu Y, Wu Y, Xu M et al. (2016) Waste electrical and electronic equipment (WEEE) recycling for a sustainable resource supply in the electronics industry in China. J Clean Prod 127:331–338. doi:10.1016/j.jclepro.2016.04.041.

116.   Guerrero LA, Maas G, Hogland W (2013) Solid waste management challenges for cities in developing countries. Waste Manag 33:220–232. doi:10.1016/j.wasman.2012.09.008

117.   Guo M, Murphy RJ (2012) LCA data quality: Sensitivity and uncertainty analysis. Sci Total Environ 435–436:230–243. doi:10.1016/j.scitotenv.2012.07.006.

118.   Gurung M, Adhikari BB, Kawakita H et al. (2013) Recovery of gold and silver from spent mobile phones by means of acidothiourea leaching followed by adsorption using biosorbent prepared from persimmon tannin. Hydrometallurgy 133:84–93. doi:10.1016/j.hydromet.2012.12.003.

119.   Ha VH, Lee J chun, Jeong J et al. (2010) Thiosulfate leaching of gold from waste mobile phones. J Hazard Mater 178:1115–1119. doi:10.1016/j.jhazmat.2010.01.099.

120.   Habashi F (2005) A short history of hydrometallurgy. Hydrometallurgy 79:15–22. doi:10.1016/j.hydromet.2004.01.008.

121.   Hadi P, Xu M, Lin CSK et al. (2015) Waste printed circuit board recycling techniques and product utilization. J Hazard Mater 283:234–243. doi:10.1016/j.jhazmat.2014.09.032

122.   Hagelüken C (2006) Improving metal returns and eco-efficiency in electronics recycling metals smelting and refining. 218–223.

123.   Hall WJ, Williams PT (2007) Separation and recovery of materials from scrap printed circuit boards. Resour Conserv Recycl 51:691–709. doi:10.1016/j.resconrec.2006.11.010.

124. Hallstedt SI, Thompson AW, Lindahl P (2013) Key elements for implementing a strategic sustainability perspective in the product innovation process. J Clean Prod 51:277–288. doi:10.1016/j.jclepro.2013.01.043.

125. Hassan NM, Rasmussen PE, Dabek-Zlotorzynska E et al. (2007) Analysis of environmental samples using microwave-assisted acid digestion and inductively coupled plasma mass spectrometry: Maximizing total element recoveries. Water Air Soil Pollut 178:323–334. doi:10.1007/s11270-006-9201-3.

126. Havlik T, Orac D, Petranikova M et al. (2010) Leaching of copper and tin from used printed circuit boards after thermal treatment. J Hazard Mater 183:866–873. doi:10.1016/j.jhazmat.2010.07.107.

127. Hawkes RB, Franzmann PD, O'hara G, Plumb JJ (2006) *Ferroplasma cupricumulans* sp. nov., a novel moderately thermophilic, acidophilic archaeon isolated from an industrial-scale chalcocite bioleach heap. Extremophiles 10:525–530. doi:10.1007/s00792-006-0527-y.

128. Hedrich S, Johnson DB (2013) Aerobic and anaerobic oxidation of hydrogen by acidophilic bacteria. FEMS Microbiol Lett 349:40–45. doi:10.1111/1574-6968.12290.

129. Hedrich S, Schlömann M, Barrie Johnson D (2011) The iron-oxidizing proteobacteria. Microbiology 157:1551–1564. doi:10.1099/mic.0.045344-0.

130. Hennebel T, Boon N, Maes S, Lenz M (2015) Biotechnologies for critical raw material recovery from primary and secondary sources: R&D priorities and future perspectives. N Biotechnol 32:121–127. doi:10.1016/j.nbt.2013.08.004.

131. Herat S, Agamuthu P (2012) E-waste: a problem or an opportunity? Review of issues, challenges and solutions in Asian countries. Waste Manag Res 30:1113–1129. doi:10.1177/0734242X12453378.

132. Herrera L, Ruiz P, Aguillon JC, Fehrmann A (1989) A new spectrophotometric method for the determination of ferrous iron in the presence of ferric iron. J Chem Technol Biotechnol 44:171–181. doi:10.1002/jctb.280440302.

133. Herva M, Roca E (2013) Review of combined approaches and multi-criteria analysis for corporate environmental evaluation. J Clean Prod 39:355–371. doi:10.1016/j.jclepro.2012.07.058.

134. Hetherington AC, Borrion AL, Griffiths OG, McManus MC (2014) Use of LCA as a development tool within early research: Challenges and issues across different sectors. Int J Life Cycle Assess 19:130–143. doi:10.1007/s11367-013-0627-8.

135. Van Hille RP, Peterson KA, Lewis AE (2005) Copper sulphide precipitation in a fluidised bed reactor. Chem Eng Sci 60:2571–2578. doi:10.1016/j.ces.2004.11.052.

136. Hong Y, Valix M (2014) Bioleaching of electronic waste using acidophilic sulfur oxidising bacteria. J Clean Prod 65:465–472. doi:10.1016/j.jclepro.2013.08.043.

137. Huang IB, Keisler J, Linkov I (2011) Multi-criteria decision analysis in environmental sciences: Ten years of applications and trends. Sci Total Environ 409:3578–3594. doi:10.1016/j.scitotenv.2011.06.022.

138. Huber G, Spinnler C, Gambacorta A, Stetter KO (1989) *Metallosphaera sedula* gen, and sp. nov. represents a new genus of aerobic, metal-mobilizing, thermoacidophilic archaebacteria. Syst Appl Microbiol 12:38–47. doi:10.1016/S0723-2020(89)80038-4.

139. Huisman J (2010) WEEE recast: from 4 kg to 65%: the compliance consequences. UNU Expert Opinion on the EU European Parliament Draft Report on the WEEE Directive with updates of the 2007 WEEE Review study and estimated kilograms per head for 2013/ 2016 for all EU27+2 countries, UNU-ISP, Bonn, Germany.

140. Huisman J, Botezatu I, Herreras L et al. (2015) Countering WEEE Illegal Trade (CWIT) Summary report, market assessment, legal analysis, crime analysis and recommendations Roadmap. Lyon, France.

141. Ilyas S, Anwar MA, Niazi SB, Afzal Ghauri M (2007) Bioleaching of metals from electronic scrap by moderately thermophilic acidophilic bacteria. Hydrometallurgy 88:180–188. doi:10.1016/j.hydromet.2007.04.007.

142. Ilyas S, Lee J (2014a) Biometallurgical Recovery of Metals from Waste Electrical and Electronic Equipment: a Review. ChemBioEng Rev 1:148–169. doi:10.1002/cben.201400001.

143. Ilyas S, Lee JC (2015) Bioprocessing of electronic scraps. Microbiol Miner Met Mater Environ 307–328.

144. Ilyas S, Lee JC, Crystal L (2015) Hybrid leaching: An emerging trend in bioprocessing of secondary resources. Microbiol Miner Met Mater Environ 359–382.

145. Ilyas S, Lee JC (2014b) Bioleaching of metals from electronic scrap in a stirred tank reactor. Hydrometallurgy 149:50–62. doi:10.1016/j.hydromet.2014.07.004.

146. Ilyas S, Ruan C, Bhatti HN et al. (2010) Column bioleaching of metals from electronic scrap. Hydrometallurgy 101:135–140. doi:10.1016/j.hydromet.2009.12.007.

147. Ishigaki T, Nakanishi A, Tateda M et al. (2005) Bioleaching of metal from municipal waste incineration fly ash using a mixed culture of sulfur-oxidizing and iron-oxidizing bacteria. Chemosphere 60:1087–1094. doi:10.1016/j.chemosphere.2004.12.060.

148. Işildar A, van de Vossenberg J, Rene ER et al. (2015) Two-step bioleaching of copper and gold from discarded printed circuit boards (PCB). Waste Manag. http://dx.doi.org/10.1016/j.wasman.2015.11.033.

149. Jadhav U, Hocheng H (2015) Hydrometallurgical recovery of metals from large printed circuit board pieces. Sci Rep 5:14574. doi:10.1038/srep14574.

150. Jadhav U, Hocheng H (2012) A review of recovery of metals from industrial waste. J Achiev Mater 54:159–167.

151. Jan RL, Wu J, Chaw SM et al. (1999) A novel species of thermoacidophilic archaeon, *Sulfolobus yangmingensis* sp. nov. Int J Syst Bacteriol 49 Pt 4:1809–1816. doi:10.1099/00207713-49-4-1809.

152. Janyasuthiwong S, Rene ER, Esposito G, Lens PNL (2015) Effect of pH on Cu, Ni and Zn removal by biogenic sulfide precipitation in an inversed fluidized bed bioreactor. Hydrometallurgy 158:94–100. doi:10.1016/j.hydromet.2015.10.009.

153. Janyasuthiwong S, Ugas R, Rene ER et al. (2016) Effect of operational parameters on the leaching efficiency and recovery of heavy metals from computer printed circuit boards. J Chem Technol Biotechnol 91:2038–2046. doi:10.1002/jctb.4798.

154. Jha MK, Choubey PK, Jha AK et al. (2012) Leaching studies for tin recovery from waste e-scrap. Waste Manag 32:1919–1925. doi:10.1016/j.wasman.2012.05.006.

155. Jing-ying L, Xiu-li X, Wen-quan L (2012) Thiourea leaching gold and silver from the printed circuit boards of waste mobile phones. Waste Manag 32:1209–1212. doi:10.1016/j.wasman.2012.01.026.

156. Joda N, Rashchi F (2012) Recovery of ultra fine grained silver and copper from PC board scraps. Sep Purif Technol 92:36–42. doi:10.1016/j.seppur.2012.03.022.

157. Johnson DB (2014) Biomining-biotechnologies for extracting and recovering metals from ores and waste materials. Curr Opin Biotechnol 30:24–31. doi:10.1016/j.copbio.2014.04.008.

158. Johnson DB, Bacelar-Nicolau P, Okibe N et al. (2009) *Ferrimicrobium acidiphilum* gen. nov., sp. nov. and *Ferrithrix thermotolerans* gen. nov., sp. nov.: Heterotrophic, iron-oxidizing, extremely acidophilic actinobacteria. Int J Syst Evol Microbiol 59:1082–1089. doi:10.1099/ijs.0.65409-0.

159. Johnson DB, Hallberg KB (2003) The microbiology of acidic mine waters. Res Microbiol 154:466–473. doi:10.1016/S0923-2508(03)00114-1.

160. Johnson DB, Du Plessis CA (2015) Biomining in reverse gear: Using bacteria to extract metals from oxidised ores. Miner Eng 75:2–5. doi:10.1016/j.mineng.2014.09.024.

161. Johnston CW, Wyatt M a, Li X et al. (2013) Gold biomineralization by a metallophore from a gold-associated microbe. Nat Chem Biol 9:241–3. doi:10.1038/nchembio.1179.

162. Jones PT, Geysen D, Tielemans Y et al. (2013) Enhanced landfill mining in view of multiple resource recovery: A critical review. J Clean Prod 55:45–55. doi:10.1016/j.jclepro.2012.05.021.

163. Jong T, Parry DL (2003) Removal of sulfate and heavy metals by sulfate reducing bacteria in short-term bench scale upflow anaerobic packed bed reactor runs. Water Res 37:3379–3389. doi:10.1016/S0043-1354(03)00165-9.

164. Jujun R, Yiming Q, Zhenming X (2014) Environment-friendly technology for recovering nonferrous metals from e-waste: Eddy current separation. Resour Conserv Recycl 87:109–116. doi:10.1016/j.resconrec.2014.03.017.

165. Kaksonen AH, Mudunuru BM, Hackl R (2014) The role of microorganisms in gold processing and recovery—A review. Hydrometallurgy 142:70–83. doi:10.1016/j.hydromet.2013.11.008.

166. Kalmykova Y, Patrício J, Rosado L, Berg PEO (2015) Out with the old, out with the new - The effect of transitions in TVs and monitors technology on consumption and WEEE generation in Sweden 1996-2014. Waste Manag 46:511–522. doi:10.1016/j.wasman.2015.08.034.

167. Kamran HH, Moradkhani D, Sedaghat B et al. (2013) Production of copper cathode from oxidized copper ores by acidic leaching and two-step precipitation followed by electrowinning. Hydrometallurgy 133:111–117. doi:10.1016/j.hydromet.2012.12.004.

168. Karak T, Bhagat RM, Bhattacharyya P (2012) Municipal solid waste generation, composition, and management: The world scenario. Crit Rev Environ Sci Technol 42:1509–1630.

169. Karwowska E, Andrzejewska-Morzuch D, Łebkowska M et al. (2014) Bioleaching of metals from printed circuit boards supported with surfactant-producing bacteria. J Hazard Mater 264:203–210. doi:10.1016/j.jhazmat.2013.11.018.

170. Kashefi K, Tor JM, Nevin KP, Lovley DR (2001) Reductive precipitation of gold by dissimilatory Fe (III)-reducing Bacteria and Archaea. Appl Environ Microbiol 67:3275–3279. doi:10.1128/AEM.67.7.3275.

171. Kasper AC, Berselli GBT, Freitas BD et al. (2011) Printed wiring boards for mobile phones: Characterization and recycling of copper. Waste Manag 31:2536–2545. doi:10.1016/j.wasman.2011.08.013.

172. Kelly DP, Wood AP (2000) Reclassification of some species of *Thiobacillus Acidithiobacillus* gen . nov ., *Halothiobacillus* gen. nov. and *Thermithiobacillus* gen. nov. Int J Syst Evol Microbiol 50:511–516.

173. Khalili NR, Duecker S (2013) Application of multi-criteria decision analysis in design of sustainable environmental management system framework. J Clean Prod 47:188–198. doi:10.1016/j.jclepro.2012.10.044.

174. Khoo KM, Ting YP (2001) Biosorption of gold by immobilized fungal biomass. Biochem Eng J 8:51–59. doi:10.1016/S1369-703X(00)00134-0.

175. Kiddee P, Naidu R, Wong MH (2013) Electronic waste management approaches: An overview. Waste Manag 33:1237–1250. doi:10.1016/j.wasman.2013.01.006.

176. Kim E, Kim M, Lee J et al. (2011a) Leaching kinetics of copper from waste printed circuit boards by electro-generated chlorine in HCl solution. Hydrometallurgy 107:124–132. doi:10.1016/j.hydromet.2011.02.009.

177. Kim E, Kim M, Lee J, Pandey BD (2011b) Selective recovery of gold from waste mobile phone PCBs by hydrometallurgical process. J Hazard Mater 198:206–15. doi:10.1016/j.jhazmat.2011.10.034.

178. Kimura S, Bryan CG, Hallberg KB, Johnson DB (2011) Biodiversity and geochemistry of an extremely acidic, low-temperature subterranean environment sustained by chemolithotrophy. Environ Microbiol 13:2092–2104. doi:10.1111/j.1462-2920.2011.02434.x.

179. Knowles CJ (1976) Microorganisms and cyanide. Bacteriol Rev 40:652–680.

180. Kolias K, Hahladakis JN, Gidarakos E (2014) Assessment of toxic metals in waste personal computers. Waste Manag 34:1480–1487. doi:10.1016/j.wasman.2014.04.020.

181. Kordosky GA (2002) Copper recovery using leach / solvent extraction / electrowinning technology: Forty years of innovation , 2.2 million tonnes of copper annually. The Journal of The South African Institute of Mining and Metallurgy: November/December 2002: 445-450.

182. Kozubal MA, Dlakić M, Macur RE, Inskeep WP (2011) Terminal oxidase diversity and function in "*Metallosphaera yellowstonensis*": Gene expression and protein modeling suggest mechanisms of Fe(II) oxidation in the *Sulfolobales*. Appl Environ Microbiol 77:1844–1853. doi:10.1128/AEM.01646-10.

183. Kumar M, Lee J, Kim M et al. (2012) Leaching of metals from waste printed circuit boards (WPCBs) using sulfuric acid and nitric acid. Environ Eng Manag J 3613.

184. Kunz D a., Nagappan O, Silva-Avalos J, Delong GT (1992) Utilization of cyanide as a nitrogenous substrate by *Pseudomonas fluorescens* NCIMB 11764: Evidence for multiple pathways of metabolic conversion. Appl Environ Microbiol 58:2022–2029.

185. Kuyucak N, Volesky B (1988) Biosorbents for recovery of metals from industrial solutions. Biotechnol Lett 10:137–142. doi:10.1007/BF01024641.

186. Ladou J, Lovegrove S (2008) Export of electronics equipment waste. Int J Occup Environ Health 14:1–10. doi:10.1179/oeh.2008.14.1.1.

187. Lee J, Pandey B (2012) Bio-processing of solid wastes and secondary resources for metal extraction–a review. Waste Manag 32:3–18. doi:10.1016/j.wasman.2011.08.010.

188. Lee S, Yoo K, Jha MK, Lee J (2015) Separation of Sn from waste Pb-free Sn–Ag–Cu solder in hydrochloric acid solution with ferric chloride. Hydrometallurgy 157:184–187. doi:10.1016/j.hydromet.2015.08.016.

189. Leff LG, Ghosh S, Johnston GP, Roberto A (2015) Microbial remediation of acid mine Drainage. Microbiol Miner Met Mater Environ 453–476.

190. Lewis A, Van Hille R (2006) An exploration into the sulphide precipitation method and its effect on metal sulphide removal. Hydrometallurgy 81:197–204. doi:10.1016/j.hydromet.2005.12.009.

191. Lewis AE (2010) Review of metal sulphide precipitation. Hydrometallurgy 104:222–234. doi:10.1016/j.hydromet.2010.06.010.

192. Li J, Gao B, Xu Z (2014a) New technology for separating resin powder and fiberglass powder from fiberglass–resin powder of waste printed circuit boards. Environ Sci Technol 48:5171–5178. doi:10.1021/es405679n.

193. Li J, Miller JD (2007) Reaction kinetics of gold dissolution in acid thiourea solution using ferric sulfate as oxidant. Hydrometallurgy 89:279–288. doi:10.1016/j.hydromet.2007.07.015.

194. Li J, Zeng X, Chen M et al. (2015) "control-Alt-Delete": Rebooting solutions for the e-waste problem. Environ Sci Technol 49:7095–7108. doi:10.1021/acs.est.5b00449.

195. Li J, Zhou Q, Xu Z (2014b) Real-time monitoring system for improving corona electrostatic separation in the process of recovering waste printed circuit boards. Waste Manag Res 32:1227–34. doi:10.1177/0734242X14554647.

196. Li L, Q H, J Z et al. (2011) Resistance and biosorption mechanism of silver ions by Bacillus cereus biomass. J Env Sci 23:108–11.

197. Li Y, Richardson JB, Niu X et al. (2009) Dynamic leaching test of personal computer components. J Hazard Mater 171:1058–1065. doi:10.1016/j.jhazmat.2009.06.113.

198. Liang CJ, Li JY, Ma CJ (2014) Review on cyanogenic bacteria for gold recovery from E-waste. Adv Mater Res 878:355–367. doi:10.4028/www.scientific.net/AMR.878.355.

199. Liang G, Mo Y, Zhou Q (2010) Novel strategies of bioleaching metals from printed circuit boards (PCBs) in mixed cultivation of two acidophiles. Enzyme Microb Technol 47:322–326. doi:10.1016/j.enzmictec.2010.08.002.

200. Liang G, Tang J, Liu W, Zhou Q (2013) Optimizing mixed culture of two acidophiles to improve copper recovery from printed circuit boards (PCBs). J Hazard Mater 250–251:238–245. doi:10.1016/j.jhazmat.2013.01.077.

201. Liu J, Mooney H, Hull V et al. (2015) Systems integration for global sustainability. Science 80, 347:963. doi:10.1016/j.cognition.2008.05.007.

202. Liu X, Tanaka M, Matsui Y (2009) Economic evaluation of optional recycling processes for waste electronic home appliances. J Clean Prod 17:53–60. doi:10.1016/j.jclepro.2008.03.005.

203. Liu Y-G, Zhou M, Zeng G-M et al. (2007) Effect of solids concentration on removal of heavy metals from mine tailings via bioleaching. J Hazard Mater 141:202–8. doi:10.1016/j.jhazmat.2006.06.113.

204. LME (2016) The London Metal Exchange, (www.lme.com), accessed 19 August 2016.

205. Luda M (2011) Recycling of printed circuit boards. Integr Waste Manag - Vol II 285–298.

206. Lundgren K (2012) The global impact of e-waste: addressing the challenge, International Labour Office, Programme on Safety and Health at Work and the Environment (SafeWork), Sectoral Activities Department (SECTOR). – Geneva: ILO, 2012.

207. Ma E, Xu Z (2013) Technological process and optimum design of organic materials vacuum pyrolysis and indium chlorinated separation from waste liquid crystal display panels. J Hazard Mater 263 Pt 2:610–7. doi:10.1016/j.jhazmat.2013.10.020.

208. Mack C, Wilhelmi B, Duncan JR, Burgess JE (2007) Biosorption of precious metals. Biotechnol Adv 25:264–271. doi:10.1016/j.biotechadv.2007.01.003.

209. Mäkinen J, Bachér J, Kaartinen T et al. (2015) The effect of flotation and parameters for bioleaching of printed circuit boards. Miner Eng 75:26–31. doi:10.1016/j.mineng.2015.01.009.

210. Mancini L, Benini L, Sala S (2016) Characterization of raw materials based on supply risk indicators for Europe. Int J Life Cycle Assess 1–13. doi:10.1007/s11367-016-1137-2.

211. Mancini L, Sala S, Recchioni M et al. (2015) Potential of life cycle assessment for supporting the management of critical raw materials. Int J Life Cycle Assess 20:100–116. doi:10.1007/s11367-014-0808-0.

212. Manzella MP, Reguera G, Kashefi K (2013) Extracellular electron transfer to fe(III) oxides by the hyperthermophilic archaeon *Geoglobus ahangari* via a direct contact mechanism. Appl Environ Microbiol 79:4694–4700. doi:10.1128/AEM.01566-13.

213. Mapelli F, Marasco R, Balloi A et al. (2012) Mineral-microbe interactions: Biotechnological potential of bioweathering. J Biotechnol 157:473–481. doi:10.1016/j.jbiotec.2011.11.013.

214. Marques AC (2013) Printed circuit boards: A review on the perspective of sustainability. J Environ Manage 131C:298–306. doi:10.1016/j.jenvman.2013.10.003.

215. Marsden J, House I (2006) The chemistry of gold extraction, 2nd ed. Society for Mining Metallurgy & Exploration.

216. Martins S (2016) Size-energy relationship in comminution, incorporating scaling laws and heat. Int J Miner Process 153:29–43. doi:10.1016/j.minpro.2016.05.020.

217. Mata YN, Torres E, Blázquez ML et al. (2009) Gold(III) biosorption and bioreduction with the brown alga *Fucus vesiculosus*. J Hazard Mater 166:612–618. doi:10.1016/j.jhazmat.2008.11.064.

218. Mateo JRSC (2012) Multi criteria analysis in the renewable energy industry. Multi-Criteria Anal Renew Energy Ind Green Energy Technol 7–10. doi:10.1007/978-1-4471-2346-0.

219. Mccann D, Wittmann A (2015) E-waste Prevention, take - back system design and policy approaches. Bonn.

220. Melaku S, Dams R, Moens L (2005) Determination of trace elements in agricultural soil samples by inductively coupled plasma-mass spectrometry: Microwave acid digestion versus aqua regia extraction. Anal Chim Acta 543:117–123. doi:10.1016/j.aca.2005.04.055.

221. Melamud VS, Pivovarova TA, Tourova TP et al. (2003) *Sulfobacillus sibiricus* sp. nov., a new moderately thermophilic bacterium. Microbiology 72:605–612. doi:10.1023/A:1026007620113.

222. Mikheenko IP, Rousset M, Dementin S, Macaskie LE (2008) Bioaccumulation of palladium by *Desulfovibrio fructosivorans* wild-type and hydrogenase-deficient strains. Appl Environ Microbiol 74:6144–6146. doi:10.1128/AEM.02538-07.

223. Mishra A, Pradhan N, Kar RN et al. (2009) Microbial recovery of uranium using native fungal strains. Hydrometallurgy 95:175–177. doi:10.1016/j.hydromet.2008.04.005.

224.   Mishra D, Rhee YH (2014) Microbial leaching of metals from solid industrial wastes. J Microbiol 52:1–7. doi:10.1007/s12275-014-3532-3.

225.   Mokone TP, van Hille RP, Lewis AE (2010) Effect of solution chemistry on particle characteristics during metal sulfide precipitation. J Colloid Interface Sci 351:10–18. doi:10.1016/j.jcis.2010.06.027.

226.   Mueller SR, Wäger PA, Widmer R, Williams ID (2015) A geological reconnaissance of electrical and electronic waste as a source for rare earth metals. Waste Manag 45:226–234. doi:10.1016/j.wasman.2015.03.038.

227.   Mukongo T, Maweja K, Ngalu BW et al. (2009) Zinc recovery from the water-jacket furnace flue dusts by leaching and electrowinning in a SEC-CCS cell. Hydrometallurgy 97:53–60. doi:10.1016/j.hydromet.2009.01.001.

228.   Natarajan G, Ramanathan T, Bharadwaj A (2015a) Bioleaching of metals from major hazardous solid Wastes. Microbiol Miner Met Mater Environ 2025:229–262.

229.   Natarajan G, Tay SB, Yew WS, Ting YP (2015b) Engineered strains enhance gold biorecovery from electronic scrap. Miner Eng 75:32–37. doi:10.1016/j.mineng.2015.01.002

230.   Natarajan G, Ting YP (2015a) Gold biorecovery from e-waste: An improved strategy through spent medium leaching with pH modification. Chemosphere 136:232–238. doi:10.1016/j.chemosphere.2015.05.046.

231.   Natarajan G, Ting YP (2014) Pretreatment of e-waste and mutation of alkali-tolerant cyanogenic bacteria promote gold biorecovery. Bioresour Technol 152:80–5. doi:10.1016/j.biortech.2013.10.108.

232.   Natarajan G, Ting YP (2015b) Gold biorecovery from e-waste: An improved strategy through spent medium leaching with pH modification. Chemosphere 136:232–238. doi:10.1016/j.chemosphere.2015.05.046.

233.   Navarro CA, von Bernath D, Jerez CA (2013) Heavy metal resistance strategies of acidophilic bacteria and their acquisition: Importance for biomining and bioremediation. Biol Res 46:363–371. doi:10.4067/S0716-97602013000400008.

234.   Navarro P, Vargas C, Alonso M, Alguacil F (2006) The adsorption of gold on activated carbon from thiosulfate-ammoniacal solutions. Gold Bull 39:93–97. doi:10.1007/BF03215535.

235.   Neilands JB (1995) Siderophores: Structure and function of microbial iron transport compounds. J. Biol. Chem. 270, 26723–26726. doi: 10.1074/jbc.270.45.26723.

236.   Nemati M, Harrison STL (2000) Comparative study on thermophilic and mesophilic biooxidation of ferrous iron. Miner Eng 13:19–24. doi:10.1016/S0892-6875(99)00146-6.

237. Neto RO, Gastineau P, Cazacliu BG et al. (2016) An economic analysis of the processing technologies in CDW recycling platforms. Waste Manag.

238. Nie H, Yang C, Zhu N et al. (2015a) Isolation of *Acidithiobacillus ferrooxidans* strain Z1 and its mechanism of bioleaching copper from waste printed circuit boards. J Chem Technol Biotechnol 90:714–721. doi:10.1002/jctb.4363.

239. Nie H, Zhu N, Cao Y et al. (2015b) Immobilization of *Acidithiobacillus ferrooxidans* on cotton gauze for the bioleaching of waste printed circuit boards. Appl Biochem Biotechnol 177:675–688. doi:10.1007/s12010-015-1772-2.

240. Niu H, Volesky B (2000) Gold-cyanide biosorption with L-cysteine. J Chem Technol Biotechnol 75:436–442.

241. Niu X, Li Y (2007) Treatment of waste printed wire boards in electronic waste for safe disposal. J Hazard Mater 145:410–416. doi:10.1016/j.jhazmat.2006.11.039.

242. Niu Z, Huang Q, Xin B et al. (2016) Optimization of bioleaching conditions for metal removal from spent zinc-manganese batteries using response surface methodology. J Chem Technol Biotechnol 91:608–617. doi:10.1002/jctb.4611.

243. Oguchi M, Sakanakura H, Terazono A (2013) Toxic metals in WEEE: Characterization and substance flow analysis in waste treatment processes. Sci Total Environ 463–464:1124–1132. doi:10.1016/j.scitotenv.2012.07.078.

244. Oguchi M, Sakanakura H, Terazono A, Takigami H (2012) Fate of metals contained in waste electrical and electronic equipment in a municipal waste treatment process. Waste Manag 32:96–103. doi:10.1016/j.wasman.2011.09.012.

245. Del Olmo A, Caramelo C, SanJose C (2003) Fluorescent complex of pyoverdin with aluminum. J Inorg Biochem 97:384–387. doi:10.1016/S0162-0134(03)00316-7.

246. Olson GJ, Brierley J a, Brierley CL (2003) Bioleaching review part B: progress in bioleaching: applications of microbial processes by the minerals industries. Appl Microbiol Biotechnol 63:249–57. doi:10.1007/s00253-003-1404-6.

247. Ongondo FO, Williams ID, Whitlock G (2015) Distinct urban mines: Exploiting secondary resources in unique anthropogenic spaces. Waste Manag 45:4–9. doi:10.1016/j.wasman.2015.05.026.

248. Orell A, Navarro CA, Arancibia R et al. (2010) Life in blue: Copper resistance mechanisms of bacteria and Archaea used in industrial biomining of minerals. Biotechnol Adv 28:839–848. doi:10.1016/j.biotechadv.2010.07.003.

249.   Oturan N, Van Hullebusch ED, Zhang H et al. (2015) Occurrence and removal of organic micropollutants in landfill leachates treated by electrochemical advanced Oxidation processes. Environ Sci Technol 49:12187–12196. doi:10.1021/acs.est.5b02809.

250.   Panda B, Das SC (2001) Electrowinning of copper from sulfate electrolyte in presence of sulfurous acid. Hydrometallurgy 59:55–67. doi:10.1016/S0304-386X(00)00140-7.

251.   Panda S, Akcil A, Pradhan N, Deveci H (2015) Current scenario of chalcopyrite bioleaching: a review on the recent advances to its heap-leach technology. Bioresour Technol 196:694–706. doi:10.1016/j.biortech.2015.08.064.

252.   Pant D, Joshi D, Upreti MK, Kotnala RK (2012) Chemical and biological extraction of metals present in E waste: A hybrid technology. Waste Manag 32:979–990. doi:10.1016/j.wasman.2011.12.002.

253.   Park J, Won SW, Mao J et al. (2010) Recovery of Pd(II) from hydrochloric solution using polyallylamine hydrochloride-modified *Escherichia coli* biomass. J Hazard Mater 181:794–800. doi:10.1016/j.jhazmat.2010.05.083.

254.   Park SI, Kwak IS, Bae MA et al. (2012) Recovery of gold as a type of porous fiber by using biosorption followed by incineration. Bioresour Technol 104:208–214. doi:10.1016/j.biortech.2011.11.018.

255.   Park YJ, Fray DJ (2009) Recovery of high purity precious metals from printed circuit boards. J Hazard Mater 164:1152–1158. doi:10.1016/j.jhazmat.2008.09.043.

256.   Pethkar A V., Kulkarni SK, Paknikar KM (2001) Comparative studies on metal biosorption by two strains of *Cladosporium cladosporioides*. Bioresour Technol 80:211–215. doi:10.1016/S0960-8524(01)00080-3.

257.   Petter PMH, Veit HM, Bernardes AM (2014) Evaluation of gold and silver leaching from printed circuit board of cellphones. Waste Manag 34:475–482. doi:10.1016/j.wasman.2013.10.032.

258.   Piccinno F, Hischier R, Seeger S, Som C (2016) From laboratory to industrial scale: a scale-up framework for chemical processes in life cycle assessment studies. J Clean Prod 135:1085–1097. doi:10.1016/j.jclepro.2016.06.164.

259.   Pilone D, Kelsall GH (2006) Prediction and measurement of multi-metal electrodeposition rates and efficiencies in aqueous acidic chloride media. Electrochim Acta 51:3802–3808. doi:10.1016/j.electacta.2005.10.045.

260.   Ping Z, ZeYun F, Jie L et al. (2009) Enhancement of leaching copper by electro-oxidation from metal powders of waste printed circuit board. J Hazard Mater 166:746–750. doi:10.1016/j.jhazmat.2008.11.129.

261. Du Plessis CA, Batty JD, Dew DW (2007) Commercial applications of thermophile bioleaching. Biomining 57–80. doi:10.1007/978-3-540-34911-2_3.

262. Plumb JJ, Gibbs G, Stott MB et al. (2002) Enrichment and characterization of thermophilic acidophiles for the bioleaching of mineral suphides. Miner Eng 15:787–794.

263. Plumb JJ, McSweeney NJ, Franzmann PD (2008a) Growth and activity of pure and mixed bioleaching strains on low grade chalcopyrite ore. Miner Eng 21:93–99. doi:10.1016/j.mineng.2007.09.007.

264. Plumb JJ, Muddle R, Franzmann PD (2008b) Effect of pH on rates of iron and sulfur oxidation by bioleaching organisms. Miner Eng 21:76–82. doi:10.1016/j.mineng.2007.08.018.

265. Pohekar SD, Ramachandran M (2004) Application of multi-criteria decision making to sustainable energy planning - A review. Renew Sustain Energy Rev 8:365–381. doi:10.1016/j.rser.2003.12.007.

266. Potysz A, Lens PNL, van de Vossenberg J et al. (2016) Comparison of Cu, Zn and Fe bioleaching from Cu-metallurgical slags in the presence of *Pseudomonas fluorescens* and *Acidithiobacillus thiooxidans*. Appl Geochemistry 68:39–52. doi:10.1016/j.apgeochem.2016.03.006.

267. Pradhan JK, Kumar S (2012) Metals bioleaching from electronic waste by *Chromobacterium violaceum* and *Pseudomonads sp*. Waste Manag Res 30:1151–1159. doi:10.1177/0734242X12437565.

268. Rawlings DE (2005) Characteristics and adaptability of iron- and sulfur-oxidizing microorganisms used for the recovery of metals from minerals and their concentrates. Microb Cell Fact 4:13. doi:10.1186/1475-2859-4-13.

269. Rawlings DE, Dew D, Du Plessis C (2003) Biomineralization of metal-containing ores and concentrates. Trends Biotechnol 21:38–44. doi:10.1016/S0167-7799(02)00004-5.

270. Rawlings DE, Johnson DB (2007) The microbiology of biomining: Development and optimization of mineral-oxidizing microbial consortia. Microbiology 153:315–324. doi:10.1099/mic.0.2006/001206-0.

271. Reck BK, Graedel TE (2012) Challenges in metal recycling. Science 80, 337:690–695. doi:10.1126/science.1217501.

272. Rees KL, Van Deventer JSJ (1999) Role of metal-cyanide species in leaching gold from a copper concentrate. Miner Eng 12:877–892. doi:10.1016/S0892-6875(99)00075-8.

273.   Reis FD, Silva AM, Cunha EC, Leão VA (2013) Application of sodium- and biogenic sulfide to the precipitation of nickel in a continuous reactor. Sep Purif Technol 120:346–353. doi:10.1016/j.seppur.2013.09.023.

274.   Reith F, Etschmann B, Grosse C et al. (2009) Mechanisms of gold biomineralization in the bacterium *Cupriavidus metallidurans*. Proc Natl Acad Sci USA 106:17757–62. doi:10.1073/pnas.0904583106.

275.   Robinson BH (2009) E-waste: An assessment of global production and environmental impacts. Sci Total Environ 408:183–191. doi:10.1016/j.scitotenv.2009.09.044.

276.   Rocchetti L, Amato A, Fonti V et al. (2015) Cross-current leaching of indium from end-of-life LCD panels. Waste Manag 42:180–187. doi:10.1016/j.wasman.2015.04.035.

277.   Rocchetti L, Vegliò F, Kopacek B, Beolchini F (2013) Environmental impact assessment of hydrometallurgical processes for metal recovery from WEEE residues using a portable prototype plant. Environ Sci Technol 47:1581–1588. doi:10.1021/es302192t.

278.   Rohwerder T, Gehrke T, Kinzler K, Sand W (2003) Bioleaching review part A: progress in bioleaching: fundamentals and mechanisms of bacterial metal sulfide oxidation. Appl Microbiol Biotechnol 63:239–48. doi:10.1007/s00253-003-1448-7.

279.   Rosner V, Wagner H-J (2012) Life cycle assessment and process development of photobiological hydrogen production: From laboratory to large scale applications. Energy Procedia 29:532–540. doi:10.1016/j.egypro.2012.09.062.

280.   Rotter VS, Ueberschaar M, Chancerel P (2013) Rückgewinnung von spurenmetallen aus elektroaltgeräten. Berlin Recycl Raw Mater Conf 481–493.

281.   Ruan J, Zhu X, Qian Y, Hu J (2014) A new strain for recovering precious metals from waste printed circuit boards. Waste Manag 34:901–907. doi:10.1016/j.wasman.2014.02.014.

282.   Rubin RS, Castro MAS De, Brandão D et al. (2014) Utilization of life cycle assessment methodology to compare two strategies for recovery of copper from printed circuit board scrap. J Clean Prod 64:297–305. doi:10.1016/j.jclepro.2013.07.051.

283.   Sahin M, Akcil A, Erust C et al. (2015) A Potential alternative for precious metal recovery from E-waste: Iodine leaching. Sep Sci Technol 150629132750004. doi:10.1080/01496395.2015.1061005.

284.   Sahinkaya E, Gungor M, Bayrakdar A et al. (2009) Separate recovery of copper and zinc from acid mine drainage using biogenic sulfide. J Hazard Mater 171:901–906. doi:10.1016/j.jhazmat.2009.06.089.

285. Sampaio RMM, Timmers RA, Kocks N et al. (2010) Zn-Ni sulfide selective precipitation: The role of supersaturation. Sep Purif Technol 74:108–118. doi:10.1016/j.seppur.2010.05.013.

286. Sampaio RMM, Timmers RA, Xu Y et al. (2009) Selective precipitation of Cu from Zn in a pS controlled continuously stirred tank reactor. J Hazard Mater 165:256–265. doi:10.1016/j.jhazmat.2008.09.117.

287. Sand W, Gehrke T, Jozsa PG, Schippers A (2001) (Bio)chemistry of bacterial leaching - direct vs. indirect bioleaching. Hydrometallurgy 59:159–175. doi:10.1016/S0304-386X(00)00180-8.

288. Sand W, Rohde K, Sobotke B, Zenneck C (1992) Evaluation of *Leptospirillum ferrooxidans* for leaching. Appl Environ Microbiol 58:85–92.

289. Schipper I, Haan (2015) Gold from children's hands: Use of child-mined gold by the electronics sector. Stichting Onderzoek Multinationale Ondernemingen (SOMO), Amsterdam.

290. Schippers A, Jozsa PG, Sand W (1996) Sulfur chemistry in bacterial leaching of pyrite. Appl Environ Microbiol 62:3424–3431.

291. Schlesinger ME, King MJ, Sole KC, Davenport WG (2011) Extractive metallurgy of copper, 5th ed. Elsevier.

292. Schluep M, Hagelueken C, Kuehr R et al. (2009) Recycling - From e-waste to resources, UNE, StEP. Berlin.

293. Schluep M, Müller E, Hilty L (2013) Insights from a decade of development cooperation in e-waste management. In: Hilty LM, Aebischer B, Andersson G, Lohmann W (eds) Proc. First Int. Conf. Inf. Commun. Technol. Sustain. pp 45–51.

294. Senanayake G (2005) Catalytic role of ammonia in the anodic oxidation of gold in copper-free thiosulfate solutions. Hydrometallurgy 77:287–293. doi:10.1016/j.hydromet.2004.12.003.

295. Senanayake G (2004) Analysis of reaction kinetics, speciation and mechanism of gold leaching and thiosulfate oxidation by ammoniacal copper(II) solutions. Hydrometallurgy 75:55–75. doi:10.1016/j.hydromet.2004.06.004.

296. Serna J, Martinez END., Narváez, PCR et al.(2016) Multi-criteria decision analysis for the selection of sustainable chemical process routes during early design stages. Chem Eng Res Des 113:28–49. doi:10.1016/j.cherd.2016.07.001.

297. Serpe A, Rigoldi A, Marras C et al. (2015) Chameleon behaviour of iodine in recovering noble-metals from WEEE: towards sustainability and "zero" waste. Green Chem 17:2208–2216. doi:10.1039/C4GC02237H.

298. Shah MB, Tipre DR, Dave SR (2014) Chemical and biological processes for multi-metal extraction from waste printed circuit boards of computers and mobile phones. Waste Manag. Res.

299. Shin D, Jeong J, Lee S et al. (2013) Evaluation of bioleaching factors on gold recovery from ore by cyanide-producing bacteria. Miner Eng 48:20–24. doi:10.1016/j.mineng.2013.03.019.

300. Silva RA, Park J, Lee E et al. (2015) Influence of bacterial adhesion on copper extraction from printed circuit boards. Sep Purif Technol 143:169–176. doi:10.1016/j.seppur.2015.01.038.

301. Silverman MP, Lundgren DG (1959) Studies on the chemoautotrophic iron bacterium *Ferrobacillus ferrooxidans*. J Bacteriol 78:326–331.

302. Simoni M, Kuhn EP, Morf LS et al. (2015) Urban mining as a contribution to the resource strategy of the Canton of Zurich. Waste Manag 45:10–21. doi:10.1016/j.wasman.2015.06.045.

303. Sliogeriene J, Turskis Z, Streimikiene D (2013) Analysis and choice of energy generation technologies: The multiple criteria assessment on the case study of Lithuania. Energy Procedia 32:11–20. doi:10.1016/j.egypro.2013.05.003.

304. Soltani A, Hewage K, Reza B, Sadiq R (2015) Multiple stakeholders in multi-criteria decision-making in the context of municipal solid waste management: A review. Waste Manag 35:318–328. doi:10.1016/j.wasman.2014.09.010.

305. Song Q, Li J (2014) Environmental effects of heavy metals derived from the e-waste recycling activities in China: A systematic review. Waste Manag 34:2587–2594. doi:10.1016/j.wasman.2014.08.012.

306. Spolaore P, Joulian C, Gouin J et al. (2011) Relationship between bioleaching performance, bacterial community structure and mineralogy in the bioleaching of a copper concentrate in stirred-tank reactors. Appl Microbiol Biotechnol 89:441–448. doi:10.1007/s00253-010-2888-5.

307. Stapleton RD, Savage DC, Sayler GS, Stacey G (1998) Biodegradation of aromatic hydrocarbons in extremely acidic environment. Appl Environ Microbiol 64:4180–4184.

308. Starosvetsky J, Zukerman U, Armon RH (2013) A simple medium modification for isolation, growth and enumeration of *Acidithiobacillus thiooxidans* (syn. *Thiobacillus*

thiooxidans) from water samples. J Microbiol Methods 92:178–182. doi:10.1016/j.mimet.2012.11.009.

309. Stenvall E, Tostar S, Boldizar A et al. (2013) An analysis of the composition and metal contamination of plastics from waste electrical and electronic equipment (WEEE). Waste Manag 33:915–922. doi:10.1016/j.wasman.2012.12.022.

310. StEP (2015) StEP E-waste World Map - European Union - STEP. In: StEP E-waste World Map.

311. Sthiannopkao S, Wong MH (2013) Handling e-waste in developed and developing countries: Initiatives, practices, and consequences. Sci Total Environ 463–464:1147–1153. doi:10.1016/j.scitotenv.2012.06.088.

312. Sun ZHI, Xiao Y, Sietsma J et al. (2015) Characterisation of metals in the electronic waste of complex mixtures of end-of-life ICT products for development of cleaner recovery technology. Waste Manag 35:227–235. doi:10.1016/j.wasman.2014.09.021.

313. Syed S (2012) Recovery of gold from secondary sources—A review. Hydrometallurgy 115–116:30–51. doi:10.1016/j.hydromet.2011.12.012.

314. Tabita R, Lundgren DG (1971) Utilization of glucose and the effect of organic compounds on the chemolithotroph *Thiobacillus ferrooxidans*. J Bacteriol 108:328–333.

315. Tanaka K, Watanabe N (2015) Study on the coordination structure of Pt sorbed on bacterial cells using X-ray absorption fine structure spectroscopy. PLoS One 10:1–12. doi:10.1371/journal.pone.0127417.

316. Tanskanen P (2013) Management and recycling of electronic waste. Acta Mater 61:1001–1011. doi:10.1016/j.actamat.2012.11.005.

317. Tay SB, Natarajan G, Rahim MNBA et al. (2013) Enhancing gold recovery from electronic waste via lixiviant metabolic engineering in *Chromobacterium violaceum*. Sci Rep 3:2236. doi:10.1038/srep02236.

318. Tecchio P, Freni P, De Benedetti B, Fenouillot F (2016) Ex-ante Life Cycle Assessment approach developed for a case study on bio-based polybutylene succinate. J Clean Prod 112:316–325. doi:10.1016/j.jclepro.2015.07.090.

319. Ting Y, Mittal AK (2002) Effect of pH on the biosorption of gold by a fungal biosorbent. Resour Environ Biotechnol 3:229–239.

320. Ting Y, Pham V (2009) Gold bioleaching of electronic waste by cyanogenic bacteria and its enhancement with bio-oxidation. Adv Mater Res 73:661–665. doi:10.4028/www.scientific.net/AMR.71-73.661.

321.   Tokuda H, Kuchar D, Mihara N et al. (2008) Study on reaction kinetics and selective precipitation of Cu, Zn, Ni and Sn with H2S in single-metal and multi-metal systems. Chemosphere 73:1448–1452. doi:10.1016/j.chemosphere.2008.07.073.

322.   Torretta V, Ragazzi M, Istrate IA, Rada EC (2013) Management of waste electrical and electronic equipment in two EU countries: A comparison. Waste Manag 33:117–122. doi:10.1016/j.wasman.2012.07.029.

323.   Tributsch H (2001) Direct versus indirect bioleaching. Hydrometallurgy 59:177–185. doi:10.1016/S0304-386X(00)00181-X.

324.   Tsydenova O, Bengtsson M (2011) Chemical hazards associated with treatment of waste electrical   and   electronic   equipment.   Waste   Manag   31:45–58. doi:10.1016/j.wasman.2010.08.014.

325.   Tue NM, Katsura K, Suzuki G et al. (2014) Dioxin-related compounds in breast milk of women from Vietnamese e-waste recycling sites: Levels, toxic equivalents and relevance of   non-dietary   exposure.   Ecotoxicol   Environ   Saf   106:220–225. doi:10.1016/j.ecoenv.2014.04.046.

326.   Tunca S, Barreiro C, Sola-Landa A et al. (2007) Transcriptional regulation of the desferrioxamine gene cluster of *Streptomyces coelicolor* is mediated by binding of DmdR1 to an iron box in the promoter of the desA gene. FEBS J 274:1110–1122. doi:10.1111/j.1742-4658.2007.05662.x.

327.   Tuncuk A, Stazi V, Akcil A et al. (2012) Aqueous metal recovery techniques from e-scrap: Hydrometallurgy   in   recycling.   Miner   Eng   25:28–37. doi:10.1016/j.mineng.2011.09.019.

328.   Tunsu C, Petranikova M, Gergorić M et al. (2015) Reclaiming rare earth elements from end-of-life products: A review of the perspectives for urban mining using hydrometallurgical   unit   operations.   Hydrometallurgy   156:239–258. doi:10.1016/j.hydromet.2015.06.007.

329.   Ubaldini S, Fornari P, Massidda R, Abbruzzese C (1998) An innovative thiourea gold leaching process. Hydrometallurgy 48:113–124. doi:10.1016/S0304-386X(97)00076-5.

330.   Ubalua A (2010) Cyanogenic glycosides and the fate of cyanide in soil. Aust J Crop Sci 4:223–237.

331.   Ueberschaar M, Rotter VS (2015) Enabling the recycling of rare earth elements through product design and trend analyses of hard disk drives. J Mater Cycles Waste Manag 17:266–281. doi:10.1007/s10163-014-0347-6.

332. Valdés J, Pedroso I, Quatrini R et al. (2008a) *Acidithiobacillus ferrooxidans* metabolism: from genome sequence to industrial applications. BMC Genomics 9:597. doi:10.1186/1471-2164-9-597.

333. Valdés J, Pedroso I, Quatrini R, Holmes DS (2008b) Comparative genome analysis of *Acidithiobacillus ferrooxidans, A. thiooxidans* and *A. caldus*: Insights into their metabolism and ecophysiology. Hydrometallurgy 94:180–184. doi:10.1016/j.hydromet.2008.05.039.

334. de Vargas I, Macaskie LE, Guibal E (2004) Biosorption of palladium and platinum by sulfate-reducing bacteria. J Chem Technol Biotechnol 79:49–56.

335. Veeken AHM, Akoto L, Hulshoff Pol LW, Weijma J (2003) Control of the sulfide (S2-) concentration for optimal zinc removal by sulfide precipitation in a continuously stirred tank reactor. Water Res 37:3709–3717. doi:10.1016/S0043-1354(03)00262-8.

336. Vegliò F, Quaresima R, Fornari P, Ubaldini S (2003) Recovery of valuable metals from electronic and galvanic industrial wastes by leaching and electrowinning. Waste Manag 23:245–252. doi:10.1016/S0956-053X(02)00157-5.

337. Vegliò F, Ubaldini S (2001) Optimisation of pure stibnite leaching conditions by Response Surface Methodology. 1:103–112.

338. Veit HM, Bernardes AM, Ferreira JZ et al. (2006) Recovery of copper from printed circuit boards scraps by mechanical processing and electrometallurgy. J Hazard Mater 137:1704–1709. doi:10.1016/j.jhazmat.2006.05.010.

339. Vera M, Schippers A, Sand W (2013) Progress in bioleaching: fundamentals and mechanisms of bacterial metal sulfide oxidation--part A. Appl Microbiol Biotechnol 97:7529–41. doi:10.1007/s00253-013-4954-2.

340. Vijayaraghavan K, Mahadevan A, Sathishkumar M et al. (2011) Biosynthesis of Au(0) from Au(III) via biosorption and bioreduction using brown marine alga *Turbinaria conoides*. Chem Eng J 167:223–227. doi:10.1016/j.cej.2010.12.027.

341. Villa-Gomez DK, van Hullebusch ED, Maestro R et al. (2014) Morphology, Mineralogy, and Solid–Liquid phase separation characteristics of Cu and Zn precipitates produced with biogenic sulfide. Environ Sci Technol 48:664–673.

342. Villares M, Işildar A, Mendoza Beltran A, Guinee J (2016) Applying an ex-ante life cycle perspective to metal recovery from e-waste using bioleaching. J Clean Prod 129:315–328. doi:10.1016/j.jclepro.2016.04.066.

343. Van de Vossenberg JLCM, Driessen AJM, Zillig W, Konings WN (1998) Bioenergetics and cytoplasmic membrane stability of the extremely acidophilic, thermophilic archaeon *Picrophilus oshimae*. Extremophiles 2:67–74. doi:10.1007/s007920050044.

344. Waksman S a., Joffe JS (1922) Microörganisms concerned in the oxidation of sulfur in the soil. J Bacteriol 7:239–256. doi:10.1097/00010694-192205000-00002.

345. Wan YZ, Wang YL, Li GJ et al. (1997) Carbon fibre felt electrodeposited by copper and its composites. J Mater Sci Lett 16:1561–1563.

346. Wang F, Huisman J, Baldé K, Stevels A (2012a) A systematic and compatible classification of WEEE. Electron. Goes Green 2012+. IEEE, Berlin, pp 1–5.

347. Wang F, Huisman J, Meskers CEM et al. (2012b) The Best-of-2-Worlds philosophy: Developing local dismantling and global infrastructure network for sustainable e-waste treatment in emerging economies. Waste Manag 32:2134–2146. doi:10.1016/j.wasman.2012.03.029.

348. Wang F, Ruediger K, Daniel A, Jinhui L (2013) E-waste in China: A country report. Bonn.

349. Wang J, Bai J, Xu J, Liang B (2009) Bioleaching of metals from printed wire boards by *Acidithiobacillus ferrooxidans* and *Acidithiobacillus thiooxidans* and their mixture. J Hazard Mater 172:1100–1105. doi:10.1016/j.jhazmat.2009.07.102.

350. Wang J, Chen C (2009) Biosorbents for heavy metals removal and their future. Biotechnol Adv 27:195–226. doi:10.1016/j.biotechadv.2008.11.002.

351. Wang R, Xu Z (2014) Recycling of non-metallic fractions from waste electrical and electronic equipment (WEEE): A review. Waste Manag 34:1455–1469. doi:10.1016/j.wasman.2014.03.004.

352. Wang X, Gaustad G (2012) Prioritizing material recovery for end-of-life printed circuit boards. Waste Manag 32:1903–1913. doi:10.1016/j.wasman.2012.05.005.

353. Wang Z, Guo S, Ye C (2016) Leaching of copper from metal powders mechanically separated from waste printed circuit boards in chloride media using hydrogen peroxide as oxidant. Procedia Environ Sci 31:917–924. doi:10.1016/j.proenv.2016.02.110.

354. Watling H (2015) Review of biohydrometallurgical metals extraction from polymetallic mineral resources. Minerals 5:1–60. doi:10.3390/min5010001.

355. Watling HR (2006) The bioleaching of sulphide minerals with emphasis on copper sulphides - A review. Hydrometallurgy 84:81–108. doi:10.1016/j.hydromet.2006.05.001.

356. Wellmer F-W, Hagelüken C (2015) The feedback control cycle of mineral supply, increase of raw material efficiency, and sustainable development. Minerals 5:527. doi:10.3390/min5040527.

357. Wendell KJ (2011) Improving enforcement of hazardous waste laws - a regional look at e-waste shipment control in Asia. Ninth Int Conf Environ Compliance Enforc 628–639.

358. Wender BA, Foley RW, Prado-Lopez V et al. (2014) Illustrating anticipatory life cycle assessment for emerging photovoltaic technologies. Environ Sci Technol 48:10531–10538. doi:10.1021/es5016923.

359. Widmer R, Oswald-Krapf H (2005) Global perspectives on e-waste. Environ Impact Assess Rev 25:436–458. doi:10.1016/j.eiar.2005.04.001.

360. Wienold J, Recknagel S, Scharf H et al. (2011) Elemental analysis of printed circuit boards considering the ROHS regulations. Waste Manag 31:530–535. doi:10.1016/j.wasman.2010.10.002.

361. Witne JY, Phillips C V. (2001) Bioleaching of Ok Tedi copper concentrate in oxygen- and carbon dioxide-enriched air. Miner Eng 14:25–48. doi:10.1016/S0892-6875(00)00158-8.

362. Won SW, Mao J, Kwak IS et al. (2010) Platinum recovery from ICP wastewater by a combined method of biosorption and incineration. Bioresour Technol 101:1135–1140. doi:10.1016/j.biortech.2009.09.056.

363. Won SW, Park J, Mao J, Yun YS (2011) Utilization of PEI-modified *Corynebacterium glutamicum* biomass for the recovery of Pd(II) in hydrochloric solution. Bioresour Technol 102:3888–3893. doi:10.1016/j.biortech.2010.11.106.

364. Xia D, Chen B, Zheng Z (2015) Relationships among circumstance pressure, green technology selection and firm performance. J Clean Prod 106:487–496. doi:10.1016/j.jclepro.2014.11.081.

365. Xiang Y, Wu P, Zhu N et al. (2010) Bioleaching of copper from waste printed circuit boards by bacterial consortium enriched from acid mine drainage. J Hazard Mater 184:812–818. doi:10.1016/j.jhazmat.2010.08.113.

366. Xiao Y, Yang Y, Van Den Berg J et al. (2013) Hydrometallurgical recovery of copper from complex mixtures of end-of-life shredded ICT products. Hydrometallurgy 140:128–134. doi:10.1016/j.hydromet.2013.09.012.

367. Xie D, Liu Y, Wu CL et al. (2003) Studies of properties on the immobilized Saccharomyces cerevisiae waste biomass adsorbing Pt^{4+}. Xiamen Univ.

368. Xie X, Xiao S, He Z et al. (2007) Microbial populations in acid mineral bioleaching systems of Tong Shankou copper mine, China. J Appl Microbiol 103:1227–1238. doi:10.1111/j.1365-2672.2007.03382.x.

369. Xu TJ, Ting YP (2009) Fungal bioleaching of incineration fly ash: Metal extraction and modeling growth kinetics. Enzyme Microb Technol 44:323–328. doi:10.1016/j.enzmictec.2009.01.006.

370.   Yamane LH, de Moraes VT, Espinosa DCR, Tenório JAS (2011) Recycling of WEEE: Characterization of spent printed circuit boards from mobile phones and computers. Waste Manag 31:2553–2558. doi:10.1016/j.wasman.2011.07.006.

371.   Yang H, Liu J, Yang J (2011) Leaching copper from shredded particles of waste printed circuit boards. J Hazard Mater 187:393–400. doi:10.1016/j.jhazmat.2011.01.051.

372.   Yang X, Sun L, Xiang J et al. (2013) Pyrolysis and dehalogenation of plastics from waste electrical and electronic equipment (WEEE): A review. Waste Manag 33:462–473. doi:10.1016/j.wasman.2012.07.025.

373.   Ye J, Yin H, Xie D et al. (2013) Copper biosorption and ions release by Stenotrophomonas maltophilia in the presence of benzo[a]pyrene. Chem Eng J 219:1–9. doi:10.1016/j.cej.2012.12.093.

374.   Yin H, Zhang X, Li X et al. (2014a) Whole-genome sequencing reveals novel insights into sulfur oxidation in the extremophile *Acidithiobacillus thiooxidans*. BMC Microbiol 14:179. doi:10.1186/1471-2180-14-179.

375.   Yin NH, Sivry Y, Avril C et al. (2014b) Bioweathering of lead blast furnace metallurgical slags by *Pseudomonas aeruginosa*. Int Biodeterior Biodegrad 86:372–381. doi:10.1016/j.ibiod.2013.10.013.

376.   Yong P, Rowson NA, Farr JPG et al. (2002) Bioreduction and biocrystallization of palladium by *Desulfovibrio desulfuricans* NCIMB 8307. Biotechnol Bioeng 80:369–379. doi:10.1002/bit.10369.

377.   Yoo JM, Jeong J, Yoo K et al. (2009) Enrichment of the metallic components from waste printed circuit boards by a mechanical separation process using a stamp mill. Waste Manag 29:1132–1137. doi:10.1016/j.wasman.2008.06.035.

378.   Young RJ, Veasey TJ (2000) Application of the ring loaded streght (RLS) disc test to monitor the effects of thermal pre-treatments on ore grindability. Miner Eng 13:783–787.

379.   Yue G, Guezennec AG, Asselin E (2016) Extended validation of an expression to predict ORP and iron chemistry: Application to complex solutions generated during the acidic leaching or bioleaching of printed circuit boards. Hydrometallurgy 164:334–342. doi:10.1016/j.hydromet.2016.06.027.

380.   Zammit CM, Cook N, Brugger J et al. (2012) The future of biotechnology for gold exploration and processing. Miner Eng 32:45–53. doi:10.1016/j.mineng.2012.03.016.

381.   Zeng X, Song Q, Li J et al. (2015) Solving e-waste problem using an integrated mobile recycling plant. J Clean Prod 90:55–59. doi:10.1016/j.jclepro.2014.10.026.

382. Zhang K, Schnoor JL, Zeng EY (2012a) E-Waste recycling: Where does it go from here? Environ Sci Technol 46:10861–10867. doi:dx.doi.org/10.1021/es303166s.

383. Zhang K, Wu Y, Wang W et al. (2015) Recycling indium from waste LCDs: A review. Resour Conserv Recycl 104:276–290. doi:10.1016/j.resconrec.2015.07.015.

384. Zhang L, Xu Z (2016) A Review of current progress of recycling technologies for Metals from Waste Electrical and Electronic Equipment. J Clean Prod 127:1–18. doi:10.1016/j.jclepro.2016.04.004.

385. Zhang X (2008) The dissolution of gold colloids in aqueous thiosulfate solutions. Murdoch University.

386. Zhang X, Yang L, Li Y et al. (2012b) Impacts of lead/zinc mining and smelting on the environment and human health in China. Environ Monit Assess 184:2261–2273. doi:10.1007/s10661-011-2115-6.

387. Zhang Y, Liu S, Xie H et al. (2012c) Current status on leaching precious metals from waste printed circuit boards. Procedia Environ Sci 16:560–568. doi:10.1016/j.proenv.2012.10.077.

388. Zhao H, Wang J, Yang C et al. (2015) Effect of redox potential on bioleaching of chalcopyrite by moderately thermophilic bacteria: An emphasis on solution compositions. Hydrometallurgy 151:141–150. doi:10.1016/j.hydromet.2014.11.009.

389. Zhou QG, Bo F, Hong Bo Z et al. (2007) Isolation of a strain of *Acidithiobacillus caldus* and its role in bioleaching of chalcopyrite. World J Microbiol Biotechnol 23:1217–1225. doi:10.1007/s11274-007-9350-6.

390. Zhou X, Guo J, Lin K et al. (2013) Leaching characteristics of heavy metals and brominated flame retardants from waste printed circuit boards. J Hazard Mater 246–247:96–102. doi:10.1016/j.jhazmat.2012.11.065.

391. Zhu N, Xiang Y, Zhang T et al. (2011) Bioleaching of metal concentrates of waste printed circuit boards by mixed culture of acidophilic bacteria. J Hazard Mater 192:614–619. doi:10.1016/j.jhazmat.2011.05.062.

392. Zimmerley SR, Wilson DG, Prater JD (1958) Cyclic leaching process employing iron oxidising bacteria. 12.

393. Zlosnik JEA, Williams HD (2004) Methods for assaying cyanide in bacterial culture supernatant. Lett Appl Microbiol 38:360–365. doi:10.1111/j.1472-765X.2004.01489.x.

D I P L O M A

For specialised PhD training

The Netherlands Research School for the
Socio-Economic and Natural Sciences of the Environment
(SENSE) declares that

Arda Işıldar

born on 2 October 1986 in Istanbul, Turkey

has successfully fulfilled all requirements of the
Educational Programme of SENSE.

Delft, 18 November 2016

the Chairman of the SENSE board

Prof. dr. Huub Rijnaarts

the SENSE Director of Education

Dr. Ad van Dommelen

The SENSE Research School has been accredited by the Royal Netherlands Academy of Arts and Sciences (KNAW)

KONINKLIJKE NEDERLANDSE
AKADEMIE VAN WETENSCHAPPEN

The SENSE Research School declares that Mr Arda Işıldar has successfully fulfilled all
requirements of the Educational PhD Programme of SENSE with a
work load of 48.9 EC, including the following activities:

SENSE PhD Courses

o   Environmental research in context (2014)
o   Research in context activity: 'Active member of the SENSE PhD Council' (2016)

Other PhD and Advanced MSc Courses

o   ETeCoS^3 PhD introductory course, University of Cassino and Southern Lazio, Italy (2014)
o   Biological treatment of solid waste, University of Cassino and Southern Lazio, Italy (2014)
o   Contaminated soils, Université Paris-Est Marne-la-Vallée, France (2015)
o   Cleaner production and sustainable management of material flows, UNESCO-IHE, The
    Netherlands (2015)
o   Environmental biotechnology, Delft University of Technology, The Netherlands (2016)
o   Contaminated sediments characterization and remediation, UNESCO-IHE, The
    Netherlands (2016)

Management and Didactic Skills Training

o   Co-organizing the UNESCO-IHE PhD week on 'Urban Sustainability', Delft (2014)
o   Assisting laboratory practicals in the MSc course 'General Microbiology' (2014)
o   Reviewer for the journals 'Scientific Reports', 'Reviews in Environmental Science and
    Bio/Technology', 'Biotechnology and Bioengineering', 'Journal of Cleaner Production',
    and 'Environment Protection Engineering' (2015-2016)

Oral Presentation

o   *Biorecovery of metals from electronic waste*, 5th International Conference on Industrial
    and Hazardous Waste Management, 27-30 September 2016, Chania (Crete), Greece

Poster Presentation

o   *Bioleaching of metals from electronic waste*, International Solid Waste Association
    World Congress (ISWA 2015), 7-9 September 2015, Antwerp, Belgium
o   *Two-step bioleaching of copper and gold from discarded printed circuit boards*,
    International Solid Waste Association World Congress (ISWA 2016), 19-21 September
    2016, Novi Sad, Serbia
o   *Cyanide production and biorecovery of gold from Pseudomonas putida*, Environmental
    impacts of mining and smelting, 8-9 January, 2015, Université Paris Sud, Orsay, France

SENSE Coordinator PhD Education

Dr. ing. Monique Gulickx

T - #0014 - 161024 - C25 - 240/170/15 - PB - 9780367087050 - Gloss Lamination